HOW HUMAN THE ANIMALS

ALSO BY THE AUTHOR:
What to Do Till the Veterinarian Comes

HOW HUMAN THE ANIMALS

By DR. JEAN POMMERY
WITH THE ASSISTANCE OF OTHILIE BAILLY

translated and with an introduction by DR. GLENN E. WEISFELD

STEIN AND DAY/*Publishers*/New York

Translated by Glenn E. Weisfeld, Ph. D.,
Department of Psychology, Wayne State University

First published in the United States of America 1979

Copyright © 1977 by Opera Mundi
Introduction and English Language Translation Copyright © 1979
by Stein and Day, Incorporated

All rights reserved.
Designed by David Miller
Printed in the United States of America
Stein and Day/*Publishers*/Scarborough House,
Briarcliff Manor, N.Y. 10510

Library of Congress Cataloging in Publication Data

Pommery, Jean.
    How Human the Animals
    Translation of Entre bêtes et hommes.
    Includes index.
    1. Animals, Habits and behavior of. 2. Animals
and civilization. I. Bailly, Othilie. II. Title.
QL751.P5313   591.5   78-24613
ISBN 0-8128-2575-6

Contents

INTRODUCTION	7
PART ONE: ANIMALS IN THEIR WORLD	17
1. Mind and Body	19
2. The Five Senses—and a Sixth?	25
3. The Dog and the Frog	36
4. The House and the City	49
5. Solitude or Family Life	67
6. Love, Eroticism, and Sexuality	85
7. Yes, They Can Speak	96
8. Play and Learning	112
PART TWO: ANIMALS AND PEOPLE	123
9. If Dogs Had Hands: Training the Animal	125
10. A Pigeon to Change the World?	132
11. People and Their Animals	140
12. The Dog and His Master	151
13. The Divine Cat	162
14. Man and His Dog	173
15. Turtles, Fish, Birds, & Co.	181
16. Man's Noblest Conquest	191
17. Childhood	196
18. The Pet and "His" Child	203
19. The Child and the Animal	211
20. Sick Animals	218
INDEX	225

HOW HUMAN THE ANIMALS

INTRODUCTION:
A New Approach to the Study of Behavior

Pets—from dogs to gerbils—significantly improve their owners' chances of recovering from heart attacks. Pets have also proved therapeutic as companions for the elderly and for severely withdrawn children.

These remarkable facts are rendered understandable by reading *How Human the Animals.* This unusual book explains how people and animals can be of mutual benefit to each other. The secret lies in our understanding animal behavior—in respecting animals' natural social tendencies and relating to them on their level.

We usually think of animal behavior as being studied in psychology laboratories. In recent years, however, the limitations of studying animals in cages have been recognized. Therefore more and more research on animal behavior is being conducted in the wild, in the species' natural habitats. The idea is to observe animals behaving not as they do under artificial conditions but in ways typical of their evolutionary past. In this way the relationship between an animal's behavior and the environment in which it evolved can be analyzed.

This approach is called *ethology:* the study of behavior from an evolutionary standpoint. The discipline has flourished in Europe for about 20 years and is catching on rapidly in North America. Many of our universities now offer psychology or biology courses in animal behavior that are taught from the ethological perspective.

The first half of *How Human the Animals* provides a clear and entertaining overview of ethology. Dr. Pommery begins by explaining, in the first chapter, that physiological factors—hormones and brain cells—shape behavior. In the second chapter he describes the senses of animals. In so doing, he demonstrates the important principle that each animal is adapted to its natural environment. This is nothing more than Darwin's theory of evolution. Thus the cat can see well at night because this animal evolved as a nocturnal hunter; the cat's vision is *adaptive* for these environmental conditions.

So the characteristics of a particular species reflect its evolutionary history. Hereditary factors—genes—shape the animal into a form that is adaptive for its particular habitat. But this is true of the animal's behavior as well as its anatomy: Animals behave in ways that suit their evolutionary past. If a particular species evolved in an environment with succulent leaves, for instance, the animal probably possesses genes not only for chewing and digesting the leaves but also for eating them—for liking their taste.

Thus there is a connection between an animal's behavior and its habitat. Behavior aids survival; conversely, survival considerations shape behavior. By studying animals in the wild, ethologists hope to understand how the behavior of a particular species aided its survival generation after generation. Good nocturnal vision helped the cat's ancestors to survive, and so the modern housecat retains this feature. We can understand the cat's superb night vision by knowing about that species' evolutionary past.

Another way to recognize the role that heredity plays in behavior is to note that genes program the brain and the endocrine (hormone-producing) system as well as the rest of the bodily tissues. Since the brain and hormones mediate behavior, genes must indirectly shape behavior. This is the underlying message of Chapter 3.

Next Dr. Pommery illustrates the extent to which animals' genetically-programmed behaviors differ. Chapter 4 describes the great variation in sociality among animals. The swan, for example, defends a territory. The wolf (the dog's ancestor) lives in a pack in which each individual occupies a definite rank in the social hierarchy. Mice enjoy socializing, but when conditions become too crowded they practice a form of birth control! Chapters 5 through 8 deal with family life, sexual behavior, communication, and play, respectively. These animal behaviors are shown to be surprisingly complex.

The second half of the book concerns interactions between people and animals. Chapter 9 discusses animal intelligence. Chapter 10 explains how man can shape an animal's behavior by taking account of its intellectual capacities and social tendencies. A key idea here is establishing dominance over the animal.

In Chapter 11 Dr. Pommery emphasizes man's moral responsibility to pets and other domesticated animals. He appeals not just to our ethical sense but also to our practical interests: We will get along best with animals if we respect them for what they are, if we "put ourselves in their paws." He pursues this theme in Chapters 12 through 16, in which he tells how to understand, care for, and appreciate dogs, cats, turtles, fish, birds, and horses.

Chapters 17 through 19 concern the child-pet relationship. Here Dr. Pommery reveals that he is concerned about the welfare of children as well as of animals. His thesis that having an animal companion can be therapeutic is quite compelling.

The final chapter constitutes a recapitulation of the underlying theme that people and animals have much in common. Dr. Pommery explains that, like ourselves, animals that are removed from their evolutionary surroundings are vulnerable to psychosomatic diseases.

The theme of the similarity between man and animals has an interesting history. Before Darwin's *Origin of Species* (1859), Western thinkers regarded man as qualitatively separate from the "beasts." In fact, 12 years later Darwin found it necessary to publish *The Descent of Man* primarily to convince skeptics that evolution applied to ourselves as well as to other species.

Today most people—certainly all reputable biologists—acknowledge that man descended from earlier primates. But a number of states still have antievolution laws on the books. And many people are reluctant to concede that human behavior—not just anatomy—is subject to the evolutionary process. This reluctance probably stems from another venerated idea, the mind-body dichotomy. It is as though man, being notoriously self-centered, wishes to keep his "mind" independent of the beasts, even if he can no longer claim full independence for his "body."

This gap that has been perceived to exist between human and animal behavior is narrowing slowly but surely. Brain researchers are making steady progress in identifying neural processes that mediate certain human behaviors and capacities, and of course most if not all of these processes have counterparts in the brains of other species. Similarly, the manifold influences of hormones on human behavior are coming to light, thereby also helping to bridge the gap. In the same vein, studies of the inheritance of various behavioral traits are revealing that even personality factors tend to run in families.

By demonstrating the complexity of animal behavior, ethology too is integrating our conceptions of man and animal. Field studies of primates in particular, notably Jane Goodall's observations of chimpanzees and the Gardners' teaching sign language to these apes (Chapter 7), have made people think twice about using an expression such as "dumb animals." Even the lowly insects, especially ants, wasps, and bees, display remarkably well-developed—and cooperative—social systems. Dr. Pommery's depiction of how dogs respond to people would probably have been dismissed as anthropomorphic ten years ago, but today it seems perfectly plausible.

It is true that animals seem to lack our degree of self-consciousness. They appear to behave "instinctively," without understanding the purposes of their actions. A bird, testing its amazing navigational system by flying south for the first time, doubtless does not know what to expect. However, many of us—including, and perhaps especially, psychologists—are only dimly aware of our own human motives.

Moreover, this distinction between man and animals on the basis of self-consciousness may well be exaggerated. If a red spot is painted on a chimp's forehead and he then sees his reflection in a mirror, he will try to feel the spot on his forehead. Most, if not all, other animals lack this level of self-awareness, but the chimp's possessing it vitiates our claim to uniqueness on this score. Similarly, a gorilla mother who is teaching her toddler to walk seems eminently human. She inches along behind him, urging him on, but ready to catch him if he falters.

Developmental psychology provides a good example of the inroads made by the biological approach. Take the phenomenon of *attachment* of the child to his mother. This concept can be traced to Konrad Lorenz's ethological studies of *imprinting*: the process by which a hatchling becomes bonded to its mother (Chapter 11). Similarly, psychologists (some of whom call themselves "human ethologists") are using ethological models to study play, aggression, altruism, parent–child interactions, communication, sexual behavior, social organization, and other aspects of human behavior.

These researchers have borrowed both the theory and the method of the animal ethologists. The theory of natural selection is invoked to help explain human behaviors. For example, mother–infant attachment is understood as an evolved adaptation to enable the helpless newborn to receive vital parental care. Similarly, ethologists speak of the adaptive value, or *survival function*, of play in permitting the rehearsal and practice of later behaviors. This functional approach has added a whole new level of explanation to behavioral science. Instead of explaining a given behavior solely in terms of its antecedents (environmental stimuli, previous learning, hormone levels, etc.), ethologists try to discover an interpretation in the light of man's particular evolutionary history—a functional explanation.

As another example, consider the fact that babies are difficult to toilet-train. We might conclude from this that babies are not very quick learners—dogs, for instance, can be paper-trained quite easily, even though their intelligence in other contexts is not especially awesome. The paradox can be resolved by noting that the evolutionary histories of the two species differ. The dog's ancestors grew up in cozy dens that had to be kept clean; consequently, puppies are models of hygiene once they are taught where the outside of their "den" is. In canines, then, continence served an essential survival function. Man, on the other hand, is a primate. Our ancestors were arboreal, and their excreta literally dropped out of sight. Consequently, an aptitude for learning continence would have had no great adaptive value and did not evolve. For this reason it is difficult to teach continence to primates, including human babies.

This example also serves to point out that each species has characteristic aptitudes and limitations. We cannot say that man is a good learner categorically—it depends on the task. Compared to the duck, man is a good mathematician but a poor swimmer and navigator. Each animal has evolved a set of sensory, perceptual, cognitive, and motor abilities that suffice to ensure survival in its natural habitat. Can we fault the subterranean mole for its poor eyesight? If so, we ought to be ashamed of our own paltry sense of smell.

The method of ethology likewise is being applied to the study of human behavior. Rather than studying people under artificial, confined conditions, human ethologists conduct *naturalistic observation*. Like an animal ethologist out in the field, these researchers unobtrusively watch and listen to people behaving spontaneously. The behaviors of interest are systematically defined and recorded, often with the aid of video or audio devices. This equipment allows the researcher to analyze behavior in fine detail and to distinguish patterns of interaction between individuals. For instance, naturalistic observation has disclosed that babies influence mothers' behavior, as well as the reverse. A mother who responds promptly to her baby's crying during the first six months is likely to have a baby who cries less (not more, as learning theory would predict) during the second six months; the baby of an unresponsive mother will cry more.

Ethologists' emphasis on the genetically based aspects of behavior is not meant as a denial of the importance of experience. Certainly, human behavior is shaped by previous learning and by environmental stimuli. But this is true of animal behavior too—some Japanese monkeys have even acquired what might be called cultural practices (Chapter 8). The point is that all behaviors—even man's capacities for learning and culture—are made possible by nervous systems that are molded by the actions of genes that have been selected by evolution. All behaviors represent the products of hereditary and environmental factors that are inextricably interwoven. No behavior is entirely learned, for learning requires the mediation of a brain; and no behavior is solely genetic, because the organism always has an environmental context and history.

Since no behavior is completely determined by genes, the possibility for modifying it always exists. This holds even for genetic defects. For example, some babies are born with a condition called phenylketonuria (PKU), characterized by mental retardation and hyperactivity. Due to a mutation, the gene for breaking down a substance called phenylalanine is absent. The excess phenylalanine accumulates in the blood and causes brain damage. If the condition is diagnosed promptly, however, it can be treated effectively. Therapy consists of eliminating from the child's diet foods that contain large amounts of phenylalanine. Thus even the serious behavioral effects of this genetic disease can be radically altered by environmental intervention. Moreover, discovery of the means of treatment was made possible by understanding the biological basis of the disorder.

Ironically, then, studying the biological basis of a trait—even a behavioral trait—can lead to effective intervention. Harry Harlow found that young rhesus monkeys raised in social isolation were pathologically timid when finally introduced to peers. He successfully treated this condition by first caging the deprived monkeys with young, nonthreatening monkeys; the latter appeared to reassure the "patients." It has been suggested that a similar treatment might work on socially deprived children, who exhibit similar symptoms.

The biological approach, then, is not intrinsically conservative; it does not imply that attempts at intervention are futile. Nor does ethology lend categorical support to any particular political ideology. Thus, some ethological descriptions of human nature have stressed man's aggressive past (e.g., Lorenz), while others have dwelt on our species' well-developed capacity for altruism (e.g., the sociobiologist Robert Trivers). Obviously, humans can be aggressive or altruistic depending upon present circumstances and past experience. In people as well as animals, antecedent events influence behavior within the bounds set by the genes. Man has sufficient anatomical and behavioral means for being both aggressive and altruistic. Obviously, the more we learn about how these biological potentials are actualized, the better we can understand ourselves and each other. Dr. Pommery's message that compatibility between man and animal is based on mutual understanding applies *a fortiori* to human relations.

This, then, is the ethologist's faith. Only by boldly facing up to our human nature can we bring out the best in ourselves. Only by recognizing man's limitations can we realize our potential. If, on the other hand, we pretend that man's behavior is infinitely flexible, we have no way of knowing how to proceed—no way of discovering what social conditions are compatible with our bodily and emotional needs. How much crowding can we tolerate, how much noise? How can we promote altruism? Are sex roles biologically based or culturally arbitrary? Can child care be adequately performed by nonrelatives? How can a parent be taught to be sensitive to the "instructions" emitted by babies to direct their own upbringing? What are the functions of guilt, pride, anger, gratitude, and jealousy? Ethology—and the related discipline of sociobiology—are shedding light on these and other vital issues. The answers appear to be complicated, but we cannot proceed rationally without them.

Let me conclude by reiterating my main point. In *How Human the Animals*, Dr. Pommery systematically shows how knowledge of animal behavior can help us to increase our enjoyment of pets. This same ethological approach is being applied to the study of human behavior. In awarding the 1973 Nobel Prize in Physiology or Medicine to three ethologists (Lorenz, Karl von Frisch, and Niko Tinbergen), the judges were endorsing this same belief: We share much in common with the animals; consequently, they have much to teach us about ourselves.

GLENN E. WEISFELD, PH. D.

PART ONE

Animals in Their World

ONE
Mind and Body

In a game preserve in France, there lives a charming doe that could easily be the heroine of a Walt Disney story.

This doe lives apart from the other animals, a hundred yards from the herd. From time to time she turns her head and stares at them, but she never approaches them.

People who observe her are astonished by her behavior. They say she acts as though she has done something terrible and is being punished. Their impression is justified; she is guilty of barrenness. She cannot have any fawns of her own.

Moreover, the herd has rejected her. As a concession, the other deer have allowed her to live a hundred yards from them; but she is strictly forbidden from crossing this space and rejoining her fellows. She knows this is so; she doesn't even try. But if she were not able to see them any more, she would die.

This animal story could be a human drama. Not so long ago, a barren Moslem woman would be scorned and dismissed from the harem.

This heart-rending story would itself have a tragic ending if the herd, instead of living in a preserve, lived in the wild. If a big cat, looking for its lunch, had come across the herd, the deer would have fled, abandoning the outcast without a moment's hesitation.

And the doe would have accepted her fate and allowed herself to be caught without a struggle. She would feel that she was valuable only as a sacrifice; she would offer herself up to the wild cat so that it would spare the others.

In ancient times, an Athenian maiden was offered to the Minotaur in order to save the city. There is never very much difference between the acts of man and those of animals.

However, this sentimental drama is primarily a scientific account. In reality, our protagonist owes her unhappiness to a hormonal problem.

Hormones are biochemical substances produced by certain glands and discharged directly into the bloodstream. These glands—the endocrine system—include the thyroid gland, the adrenal glands, the gonads (ovaries and testes), the pancreas, and the pituitary (located at the base of the brain and an enigma for a long time).

The endocrine system resembles a republic, with the pituitary as president and the other glands as cabinet officers. These glands seldom act alone; they usually act in concert. When one of them secretes too much of a hormone, the pituitary reacts at once, releasing its own hormones to neutralize the excesses of its subordinate.

Obviously, anything that involves the endocrine system is extremely important. Thus, adrenalin, produced by the adrenal glands, is responsible for the nervousness and split-second reactions of cats. The adrenals are so well developed in this species that a young veterinarian who was trying to remove a cat's ovary mistook an adrenal gland for it instead. Fortunately he did not remove the other adrenal too, which would have been fatal.

The entire reproductive cycle depends on hormonal regulation. A lioness allows the male to approach her only when she is in heat. And she is in heat only when the ovary has secreted a hormone: estrogen.

The gonadal hormones affect the entire organism. Men and

roosters both react to castration by becoming fatter; and men lose their masculine hairiness, whereas roosters lose the brilliant colors of their feathers. In both species, the voice gets higher. As for hens that are deprived of their female glands, they become like roosters—spurs emerge from their feet!

One yokes oxen, which are gentle and patient, not bulls. If you castrate a mean dog, he loses his aggressiveness.

Aggressiveness is promoted by the male hormone, testosterone, which lends the animal strength and grace. We shall see how testosterone can affect behavior.

The principal female hormone that affects sexual behavior is estrogen. A different hormone, also produced by the ovary, influences parental behavior. This is progesterone, which induces a bird to build its nest, produces crop milk in the pigeon, and promotes pregnancy and lactation in female mammals.

A little too much, or not quite enough, of one of these hormones will transform an animal anatomically and behaviorally.

Hormones are sometimes responsible for dramas such as that of our deer heroine. They can also have comic results. A bitch whose mammary tumor was treated with testosterone began to lift her leg and to fight with dogs! A very old dog who received estrogen treatment for his prostate began urinating like a female!

When the importance and far-reaching effects of hormones on behavior were revealed by research, an entirely new aspect of animal life came to light. Thanks to hormone research, scientific explanations of many previously incomprehensible facts became possible.

The zoologist Georges Cuvier (1769–1832) was the first to believe that the animal kingdom could not be understood without first studying anatomy and biology, which were the bases of his research. Subsequently, biological discoveries have established—and this was inconceivable 150 years ago—that all families of living things are related to each other.

Today we believe that the mind is inextricably related to

the body and expresses itself—through gestures, feelings, language—via the body's mediation. This is true of animals and also of man, who—*pace* Descartes!—although superior, is nevertheless a mammal like the cat, the dog, and the dolphin. No less astonishing is the similarity between animals and man in the realm of behavior. As we shall see, the similarity extends far beyond such obvious resemblances as that of a crowd to an anthill!

The animal kingdom is divided into vertebrates and invertebrates, which are themselves subdivided into classes, orders, families, and species. All animals have a common biological element: the cell. Fish, man, mammal, bird—every living creature is composed of cells.

Cells are found in the blood, in the form of red and white corpuscles.

They are similar functionally from species to species, but their size differs. The smaller the animal, the larger the red blood cells.

Those of the camel are the most remarkable. First of all, they are oval, not round. Furthermore, they are the only red blood cells known that can survive in water—those of all other animals burst and die.

One might suppose that this peculiarity is due to the fact that the camel can, after several days of deprivation, swallow five or six gallons of water without ill effects—an intake that would cause fatal hemorrhaging in any other animal. But let us turn things around. It is really the peculiarity of the cell that allows the camel to drink enormous quantities of water.

All animals, including man, obey the same laws of life. They breathe, they feed themselves, they discharge wastes. Blockage of any one of these basic functions brings about death.

And all of them, from the tiniest insect to the elephant, possess a "survival instinct" that shapes their behavior, giving rise to a constant struggle for life from birth to the end. All of them reproduce, feed, and defend themselves against predators. These instincts have been called "biological intelligence." As we shall see, the problems that this intelligence sometimes poses resist the efforts of the best biologists.

Where does this biological intelligence originate? Obviously, in the brain. But, in addition, the poorly understood domain of hormones is also involved, which is why the term "instinct" is applied.

Let us skip briefly over the invertebrates, which are regarded as inferior animals. Their nervous system is diffused throughout the body. However, in the past few years it has been discovered that insects and crustaceans also possess an endocrine system.

By way of contrast, all the vertebrates possess a brain. However, the brain of a primitive beast—a relic of prehistory—is enormously different from that of man or any other higher animal such as the horse or cat.

The alligator that, with eyes that appear to be closed, pretends to be asleep is a living trap ready to snatch its prey. Its brain is fashioned for this simple existence. It does not know fear. In 220 million years, the alligator has not progressed. Its brain is a long mass (because of the nose) with olfactory lobes and well-developed eyes—American crocodiles have very sharp vision. The cerebellum is unremarkable. The cerebrum is small and lacks convolutions.

The cerebellum and the cerebral hemispheres of man's brain are covered with convolutions. But before we start boasting about this feature, we should bear in mind that convoluted brains are also found in the horse, which is no intellectual giant, and the sheep, a bleating moron.

For many years it was believed that intelligence is measured by the weight of the brain. A heavy brain was considered superior to a light one. This notion is still widely accepted.

The tortoise approaches the record for being small-brained. Its brain is only 1/2240 of its total weight. Thus we should expect its sluggishness not to be confined to its feet. The elephant clearly surpasses the tortoise; it has 11 to 13 pounds of brains. But this is a mere trifle compared to its seven tons of flesh!

Obviously it is man, with a brain mass that varies between 1/30 and 1/35 of its weight (an average of 2.75 pounds for a height of five feet seven inches), that takes the prize. The most surprising is the dolphin, a very strange animal with a brain of about 3.75 pounds in a seven-foot body.

Because the cat's brain represents 1/8 of its weight and that of the lion only 1/500, we might wonder whether nature made a mistake somewhere. The big feline does not seem to be less intelligent than its little brother.

The truth—today's truth, anyway—is that the animal's degree of advancement equals the complexity plus the weight of its brain. Therefore, if one equates adaptation with evolution, the "civilized," domesticated cat should be more highly evolved than its wild brother.

TWO
The Five Senses— and a Sixth?

Atlas was one of the most famous detectives in Paris. A police dog, he was known to be capable of following a trail for miles to recover a lost child lying exhausted at the foot of a tree. What human could match these talents?

Atlas could do all this, and he carried out his duties with such ease and élan that officers asked for his assistance at the Paris police department, where he was highly respected. It was indeed an honor to receive a visit from him, to see his large, calm, reassuring form as he pushed open the office door with his muzzle. Atlas enjoyed all the respect due a chief inspector.

The police officers still recount his exploits with pride and emotion. His master once said of him, "That animal has such an incredible sense of smell! He must be a thousand times better than a man." But he quickly added, as though he had betrayed great personal vanity, "Well, maybe I'm exaggerating a bit."

In actuality, his estimate was overly modest. Atlas's sense of smell was a million times more powerful than the officer had imagined.

Release a dog in a forest. He will immediately mark his territory by urinating on trees. A year later, lead him to the same spot. You will see him romp joyfully from tree to tree. He recognizes them all—not by sight, because he is nearsighted, but by smell.

With his particularly sharp sense of smell, a hunting dog can trace the scent of a rabbit among the thousand odors of a forest. He can follow the trail of "his" quarry as if it were marked in red. Even if other animals pass that way and add their confusing odors, and in spite of any zigzagging that the prey may attempt in order to lose its pursuer, he will not lose his way. To him, there are as many distinct odors as there are rabbits.

It is because of their delicate sense of smell that man has always made use of "lower" animals for delicate, even dangerous, jobs such as mine-clearing.

Several years may go by after the end of a war before the authorities decide to clear a mine-field. Nevertheless, the dog will find all of the mines. Sometimes a mine will explode, but only rarely—thanks to God's compassion. Actually it isn't the mine at all that the dog searches for. Wherever the soil is disturbed, a particular oxidation reaction is produced; it is the resulting odor that the animal responds to—sometimes four or five years later!

The animal with this extraordinary sense of smell is myopic from birth. He sees as if in a fog—not farther than 100 yards under the best of conditions. The dog "sees" with his nose. If he races toward his master, it isn't because he recognizes him by sight but because the wind has carried his master's scent in his direction.

Frequently at around 12 years of age a dog acquires cataracts that blind him. But blindness does not interfere with normal life—especially if the dog's master has the wisdom to leave him alone in his own territory in the yard or street. The dog will find his own scent-marks and will learn to get about.

This is how the blind shepherd dog—the little Pyrenean breed, with the ruffled hair—continues to guard his herd of cows. He "senses"—the term is appropriate—the animal that is wandering away. And he brings it back by biting the rear leg, just as other shepherd dogs do. Without ever making a mistake, he nips the hind leg, not the foreleg. This is proof positive that the legs of a cow have different smells.

A cat's life is tied to its eyes. We find one-eyed cats, but blind ones rarely survive. A cat without eyes is a doomed animal.

Although dogs distinguish colors fairly well, cats essentially see in black and white, although with extraordinary acuity.

Consider the cat. His lowered eyelid allows only a sliver of light to enter, he has rolled himself into a ball, he appears oblivious to the world, he seems to be asleep. In reality, he is at his observation post. Who hasn't had the experience of sneaking up on a motionless cat and grabbing nothing but air?

The cat's eye has very little in common with man's. It is a sort of objective lens with several focal lengths and a highly sensitive photocell. The diaphragm need be opened only 1/16 for the image to be captured with perfect clarity.

At night the pupil dilates to such an extent that the iris is no longer visible. A special pigment at the back of the eye functions like highly sensitive film. Despite the darkness, the image is distinct. The eye traps the slightest luminosity.

The cat has what photographers call a "wide angle" of binocular vision. It subtends 130°, which is greater than man's 125° and the dog's 85°.

He can also transform his eye into a sort of telephoto lens. Watch him on a balcony as he trembles with such excitement that it is obvious that he has spied a pigeon. He crouches, gets ready to leap, jumps—and plunges to the ground. Caught up in the excitement, he forgot to switch to normal vision. With his eyes trained on the bird, he could see nothing else. Everything else was blurry.

Cats have always seemed to man to be more mysterious than dogs. This may be due in part to the cat's whiskers. To reassure ourselves, we classify them as part of his sense of touch, but is this accurate?

These are the vibratory bristles of the cat. They adorn his ears, constitute his eyelashes, and above all form the magnificent Dali-esque mustache that allows him to walk across a table covered with fragile ornaments without breaking a single one.

The same organs permit the leopard to avoid running into trees when he hunts. A predator that ambushes his prey just like a cat lying in wait for a bird, the leopard hurtles along with abandon, just as long as his ample mustache—which is usually as wide as his body—transmits no message of a pitfall along the way. (Insects also possess sensitive bristles: antennae that bear more of a resemblance to miniature radar units than to the usual sense organ.)

Felines and canines both have ears eight to ten times better than man's, and mobile in the bargain. On the other hand, their sense of taste is poorly developed—although we do not know if the lion prefers a piece of venison to a gazelle leg! However, domestic animals can develop "good taste"; if their master is a gourmet, they will acquire his fastidiousness. This is called evolution, a virtue turned into a vice.

Much less intelligent than canines or felines are the ruminants, which are very different from man. One seldom commits the error of anthropomorphism with the cow!

Their senses are no more interesting than they themselves. We should note, however, the importance of the nose in their social, alimentary, and sexual behavior. Thus a bull—a powerful male if ever there was one—can maintain an erection only if he has sniffed the object of his desire for a good half hour.

The horse, too, is full of curiosities. His eyes are like bifocals. When a horse lowers his head, he sees simultaneously the tender, green grass that he loves and, in the distance, the man who approaches him (or, in the case of a zebra, the lion in search of its dinner).

Another interesting feature is the horse's hoof, which has a very sensitive sense of touch and allows him to test the ground. It is very rare for a horse to sink into a bog.

The horse's sense of smell is very acute, but his taste is something else again—he loves oats and even straw!

His large ears hear very well, but not as well as the dog's. We know that the dog perceives ultrasound and that the quality of his hearing makes us appear to be half-deaf. The dog's ear, however, is practically pathetic compared with that of the dolphin, which can detect a pitch of 120,000 oscillations per second, compared to man's 25,000.

We have entered here into a realm that man can reach only with scientific instruments. But one thing is certain: Animals have five senses—at least! And even if some of these senses are reduced, they remain in a latent state.

Let's look at the fish. His right eye sees on the right side, and his left eye on the left. This is what is called monocular vision. In no way does it prevent this knife blade of an animal from having excellent vision.

The creatures that inhabit the ocean depths or subterranean rivers are blind. However, it has been learned that the internal ears that they possess are sensitive enough to compensate in part for their absence of sight. It is known too that relatives of these fish that live instead in illuminated water have normal eyes.

Lamarck said, "Function shapes an organ." The statement may be turned around: An organ ceases to exist when it no longer is functional. Thus, consider the mole: What purpose would the eyes of this fine subterranean serve?

It has been noticed that there are almost always one or two senses that are more developed than the others in every animal, and that these are the senses that allow it to survive, or at least to live longer.

Thus it appears that the senses of taste and smell of the fish actually form one and the same sense. When feeding, this animal tastes and smells the particles suspended in the water at the same time. The fish's sensitivity to the taste of salt is 180 times better than man's, and for sugar 500 times. Who would have suspected that the fish had a sweet tooth?

Fish also possess a "sixth sense." Their fish's sense of touch is lightly distributed over the entire body but, like the cat's, is something of a mystery. This sense comes from the fish's "lateral line." We know of its existence. We can verify its effects. But science still cannot explain how it works.

The lateral line organ is essential to life for the fish. It tells him just what is going on around and under him. It allows him to detect sounds, odors, and the presence of a conspecific (a member of the same species). It can also take physical measurements: of pressure and, in certain cases, the flux of pressure waves produced by its own movements.

And should we number among its senses the electrical discharges of the torpedo fish?

Of all the animals of the sea, the mammals are the most developed and the least understood by us. For example, the remarkable hearing of the dolphin is matched almost perfectly by the whale and the other mammals that inhabit the world of ultrasound, a realm unknown to man but where many species roam.

There is something more extraordinary still: the "sixth sense" of the dolphin. This sense is similar, but infinitely superior to, the sonar with which modern navies are equipped. It is so precise that it differentiates between two fish that are the same size but belong to different species! This sense is not related to olfaction, for the dolphin cannot smell, or to vision, for he sees only poorly.

The U.S. Navy has been especially interested in these astonishing creatures, wanting to train them for submarine warfare. At the undersea laboratories of the University of Florida, they have set up some fascinating experiments.

After converting the dolphins' tank into a labyrinth strewn with obstacles, the experimenters tried to jam their "sonar" transmissions by diffusing signals of the same frequencies as those emitted by the animals. Next they tossed into the tank a certain fish that dolphins love. Ignoring the interfering signals completely—and making sport of the obstacles along the way—the dolphins caught up with their food just as rapidly as usual. The scientists who set up this game—which amused the dolphins greatly—calculated that a man would have required twice the time to extricate himself from the same labyrinth, and that the interference would have thrown off the most sophisticated scientific apparatus.

Let us consider another animal, a bird that is not a bird: the bat. This nocturnal creature directs itself toward its prey—tiny insects—by using a sonar system with frequencies that are inaudible to us. The bat is a sort of guided missile.

When a caged canary tilts his head in that distinctive, almost personal, way, it isn't in order to amuse his mistress as she imagines, but rather to spot grain at his feet. "Grand vistas" do not exist for him.

Like the fish, most birds have monocular vision. Perspective and relief are unknown to them. Nevertheless their vision is excellent; their acuity is two or three times more developed than ours.

Three hundred yards above, the vulture cruising in the sky can clearly distinguish the carrion that he eats. With his own infallible eye, the eagle spots a rabbit and lands on him before the rabbit can even think about hiding.

Birds have internal ears. Their hearing is particularly well developed. It ranges from 200 to 1000 oscillations per second and even beyond in certain nocturnal species, such as the scops and the great horned owl, that have mobile, well-formed external ears.

It might be said that birds, too, have a sixth sense. This

capacity is manifested by the bird's need to sing. Everyone should listen at night to the "singing lessons" given to young nightingales, the exercises that their tutor makes them repeat time after time. This highly diverse language varies from chirping to singing. It is largely instinctual (and doubtless inferior to the visual system), being extremely important for reproduction, communication, and territorial marking.

Their melodious language compensates birds for having such a weak olfactory sense and hardly much more of a gustatory sense. Whether they eat grain or carrion, their language is not devoted to the discussion of cuisine!

One makes do with what one has. Birds have a beak and feet to use as their sense of touch. One has to make the best use of the gifts that God has parcelled out, especially when He has given you wings like those of angels!

Birds are the inhabitants of the sky. The dwellers beneath the earth are the rodents. Among the latter is found one of the most intelligent of nature's creatures: the rat.

When, in 1970, it was decided to transfer Les Halles, the old market area of Paris, to Rungis, another relocation took place underground: two million rats also decided to pull up stakes.

Two million rats left, packing their bags well in advance of the humans. Mile after mile of them marched along, with gray coats, burning eyes, and a toddling pace. The babies pressed close to their parents. The elderly died along the way from fatigue. It was an unprecedented migration, totally invisible to man.

When the first Les Halles truck arrived at Rungis, they were already installed, waiting impatiently for food.

It is likely that their delicate sense of smell leads them to their dinner, for rats are epicures with a highly developed palate. But what mysterious sense apprised them of the departure of their provisioners *before* the latter left the city? This is one of the enigmas with which these undoubtedly intelligent subterranean creatures challenge us.

"Hippopotamus" means "river horse," but this animal is more like a large pig. He is a survivor of the Tertiary Period, like the rhinoceros and the elephant. However, the elephant has good eyes and his other senses are well developed, so he is probably a little ashamed of his distant relatives.

Hippos hardly see at all, distinguishing moving shadows at most. They do not hear well either, and have to move their large heads in order to localize sounds. They are the idiots of the family! But happily for them, these poor old monsters have an extremely well developed olfactory sense.

In this area, it should be acknowledged, man is among the least favored of the animals. The lazy pig has not passed on his sense of smell to man. But wait! The pig does resemble us—in his sense of touch, which is comparable to ours, and his hearing, and the oily nature of his skin (the dermis and epidermis respond in the same way).

Pigs were released for the first atomic experiments on Bikini, in order to learn what sort of radiation burns would be inflicted on humans. And, as some good often comes out of evil, nowadays a person who suffers a large burn receives skin grafts not from a self-sacrificing relative but from a pig.

Ancient evolutionary periods have not only bequeathed to us their elephants, their rhinoceroses, and their hippopotamuses. Some other witnesses of prehistory—even, one might say, of a terrestrial paradise—the snakes and their cousins, lizards and crocodiles, seem to be on the road to extinction; only a few survivors remain.

The reptiles have a set of primitive senses plus a very peculiar sense that allows them to detect variations in temperature on the order of 1° Centigrade (1° Centigrade suffices to melt 1 gram of ice).

There once was a charming, tame grass snake—she really was tame—named Cleopatra. Her mistress was a young girl whom she loved very much. To show her affection, the snake looped herself around her mistress's neck or wrist. Anyway, the girl interpreted this as a sign of affection.

From time to time, the child took Cleopatra to visit her playmates. With some of the children, Cleopatra coiled herself around them, but she merely fell to the ground with others.

It was a matter of the child's temperature! Her mistress's temperature suited the snake perfectly, and the same with some of the other children. But with the remainder of the children, no.

Unlike fish, aquatic reptiles and lizards can see as well on land as under water.

Crocodiles and their relatives have good eyes and good teeth. Snakes also have good ears—snake charmers demonstrate that. But can we attribute this to the snake's ears, since it is the entire body that perceives vibrations?

A Mesozoic animal, the reptile actually is not very highly evolved. Nevertheless, it is more advanced than the amphibians, another throwback to prehistory.

If man (who, as cruel as he may be, remains a sentimentalist) knew that frogs weep, would he renounce his habit of eating their delicious legs? Yes, frogs cry! Without this humidification, their eyes would dry out. . . .

What else can be said of this poor creature? Well, he doubtless perceives color, since fishermen catch him with bits of red rag. But the fishermen are not that clever: Scientists have confirmed that the frog actually prefers blue!

Let us set, side by side, the skeleton of a gorilla and the skeleton of a man. This sight gives us pause. The proportions are different, of course, but the resemblance is enough to take us down a peg or two.

We observe that the cranial vault of the ape is clearly

inferior to man's, and this reassures us. But Darwin was truly not mistaken when he argued that the two forms had a common ancestor.

What remains unsolved is how the prehistoric primates (which evolved 300 million years before our era) came to be divided, 29 million years later, into two branches, one of which would become the anthropoid apes, the precursors of Homo sapiens.

In order to convince ourselves further that the ape is really our cousin, let us examine:

His sense of smell: He has no more of one than man, who has the poorest olfactory sense of all the vertebrates.

His hearing: He can hear a little better than man, but less well than the dog.

His sense of taste: Originally, poor; but did Neolithic Man have such a delicate palate? Leaving aside that epoch, let us consider the peasant of barely a hundred years ago subsisting on cabbage soup—would he have appreciated caviar? Apes, as soon as they are contacted by civilization, cultivate their taste exactly as man does.

His sense of touch: It resembles man's, being especially well developed in the hands and feet.

His vision: It is comparable to ours. The systems of adaptation and convergence are the same. Apes see very well. They perceive color. But the most surprising thing is that, for them too, the eye is an erogenous zone! When an ape sees a couple next to him making love, he immediately gets amorous ideas. (On the other hand, you could lead a bull in front of ten cows, all in the act of being serviced, and this would leave him totally cold). If one were to teach stripteasing to female apes, no doubt they would make a great success of it!

THREE
The Dog and the Frog

Youka was a handsome German shepherd with a black and tan coat. Her intelligence left an indelible memory on all who knew her. I mentioned her several times in *What to Do Till the Veterinarian Comes* (cloth, Chilton Book Company, 1977; paperback, Ballantine Books, 1977), recounting the extraordinary story of her confinement, among other incidents.

Youka selected her mate herself; the first eligible bachelor that was presented to her—a champion no less—did not please her. Anyway, she waited contentedly for the time when she would send her young into the world. Although she was seven years old, when the time came she gave birth to her first litter perfectly calmly, licking the puppies normally, as they were born, and cutting the umbilical cord.

Toward the third baby, however, she clearly registered stupefaction: the darling little pup was completely white!

The next day a red spot of blood marred the coat of hair. This would not have been noticed, amid seven blind puppies jostling and scratching each other in a basket, if it had not come into view when Youka repulsed that puppy at feeding time.

Several minutes later, yelping pierced the air—and a second, more prominent red spot appeared on the little dog's abdomen. This time the bite was so clean that there could be no doubt about it: Youka had bitten her child.

At once, mother and daughter were separated. Confined to the healing warmth of a large Persian cat, the puppy was bottle-fed and ensconced in a room on the floor above. The plan was to shelter her from the homicidal fury of the big black dog.

But Youka was able to foil the surveillance. Leaving her young, she sneaked noiselessly up the stairs and through the corridor, and opened the door—a trick she had previously mastered. She proceeded to her nursling, grabbed it from between the cat's paws, and—neatly, without letting it suffer—killed it.

She then returned to the basket and subjected two other puppies to the same fate. One puppy was malformed, and the other had muscular tremors.

This triple infanticide seems atrocious to human eyes. Actually, by killing the three puppies that she considered as not fulfilling the criteria of normality of her species, Youka had shown great wisdom: she had assured the purity of her descendants. She had acted not by intelligence, for her genius did not extend that far, but by instinct: by obeying an order given by her genetic memory.

All wild animals do the same thing.

Genetics is a relatively recent field, the discovery of which we owe to Mendel, whose research on his famous peas, conducted in 1865, was not recognized until 30 years later. But it is Jean Rostand who made the critical step in this field. One day this great scientist pondered over an unusual amphibian—a white tree frog—and out of this specimen was born modern genetics.

This white tree frog intrigued Rostand all the more because it had been born of two completely normal frogs. Actually it had only obeyed Mendel's laws. Its grandfather had been the usual gray-green shade of tree frogs. But its grandmother had been white. Skipping a generation, the trait of albinism had reappeared. The fact of the frog's color did not detract in any way from its racial purity, or from its resembling its forebear closely.

Rostand's work, together with that of other scientists, demonstrated the presence in the reproductive cells of chromosomes, whose existence was unknown previously.

Females have two x chromosomes. Males have one x chromosome and one y chromosome. When a reproductive cell divides in half before fertilization, two cells with one x chromosome each are formed in the mother and one cell with an x chromosome and another with a y chromosome are formed in the father. If a male cell with an x chromosome fertilizes the mother's cell, a daughter will be born. A son will be born if a male cell with a y chromosome wins the competition, which can be observed under the microsope, the rush of the spermatozoa giving one the impression of a race that each is trying to win.

The result of fertilization is that the sex and the species are determined by the chromosomes. A chromosome resembles a little modern sculpture that is provided with wings turning in all directions. It is a sort of extraordinary lottery of life. At a fair, one wins a teddy bear or a doll, depending upon whether the red or the green comes up. But the player never knows what the roulette wheel will reveal. When a scientist observes the mysterious world of the chromosomes, he too does not know "where she stops."

Within the chromosomes are found the much-heralded "genes": microscopic bands of nucleic acids that have been programmed in the same manner for millennia. These genes are reproduced in each child, stereotypically, as long as nothing modifies them. These bands function just like information bits in a computer. This is why the alteration of a single atom in a nucleic acid can suffice to change the morphology of the future living being.

Various factors can be the cause of such a modification. One possibility is unusually intense cosmic rays. Another is that man himself has played the sorcerer's apprentice—for example, by breeding cattle to produce more meat.

Be that as it may, the chromosome passes on the new information slavishly, like a machine. But what initially is only an aberration may be passed along from gene to gene and become, through the centuries, a permanent mutation.

In the case of random mutations, the result can be the situation in a certain Spanish village where all the inhabitants have six fingers. Alternatively, we have a village in Vendée, where people are born automatically—and that is the right term—with a harelip or with a characteristic arthrosis of the hip.

In the case of human intervention, man has so wished to breed certain races of dog that a gene for hip arthrosis has been accidentally introduced. That is why an increasing number of dogs suffer from congenital hip dysplasia.

There are also some anomalies that have existed all along but that only modern genetics can explain: white cats with blue eyes that are usually deaf; calico cats that are sterile; and those black, fiery dogs that are more aggressive than the yellow variety.

Animals, especially wild animals, possess genetic mechanisms for eliminating genes that would weaken the species. This explains why an animal may without hesitation kill one or several of its young. In so doing, she is defending her descendants against a weakness transmitted by a harmful mutation, or at least a perceived weakness.

The black panther, which is an abnormal mutant of the ordinary panther, is rejected by its conspecifics exactly as would be an albino. Rejection can, in fact, proceed to the point of assassination! Thus white blackbirds—also products of a genetic mutation—are rare only because the others suppress them.

Genetics governs the entire organization of the animal world. It determines what we refer to as an animal's "race." A Yorkshire terrier is genetically a Yorkshire terrier because he has all of the proper characteristics. But, in order to sire another Yorkshire terrier he will have to mate with a member of that same breed. If not, a pure Yorkshire terrier will not be produced.

Things can be carried even further. A tiger will never copulate with a lion. But several years ago, at the zoo of the king of Morocco, a union of this type was induced—it made quite a story—and "tigrons" were born. These tigrons could not themselves reproduce. But if, somehow, they had succeeded, the result would have been either a tiger or a lion, not a tigron.

Similarly, the hinny—the offspring of a female donkey and a horse—is not able to reproduce. And a mule must be inseminated by a donkey or by a horse in order to have young, because two mules together are unable to reproduce. This too is a matter of genetics.

Actually animals are much more racist than men, but they express their racism on the level of reproduction. That is, their racism is genetically programmed.

"Sonny," a magnificent 90-pound Airedale, did not like any other dogs, no matter what their race or sex. However, whenever nature "called," he would think nothing of running 20 miles to be near a bitch in heat.

The sole exception to his "canine phobia" was a charming, 12-pound dachshund named Toupie. Star-struck with admiration, he adored her. And Toupie could lead him around by the nose.

Naturally when Toupie was in heat the two lovers were separated. But the dogs evaded human surveillance and were able to rendezvous again. Their owners were astonished to see, however, that Sonny was not at all aroused by Toupie. It was as though he could not smell her even though she was all ready for sex. Consequently the dachshund did not show any emotion. The two dogs acted as though they recognized that nature had made them dissimilar. Or perhaps platonic love can exist—among canines, anyway.

This is not an exceptional case. It is even commonplace. Thus one cannot begin to speak of two animals' belonging to the same species unless they can and *want* to reproduce together.

Although it was born, in a manner of speaking, of a frog, genetics has not yet accomplished its metamorphosis. We do not know what it will turn into. The available knowledge allows us only to make some predictions, with all the risks that such speculations carry.

At this point, we know with certainty that genetics is the primary cause of what Darwin called "natural selection." In order to live in a particular environment, or "ecological niche," an animal must be genetically adapted to it. The animal's innate behavior is peculiarly suited to that environment, thereby assuring the survival of the species.

This law implies that the genetic inheritance of animals of the same, identical species differs according to the environments they inhabit. A French pointer would not survive an hour in the climate of the North Pole, whereas Samoyeds find the area very much to their liking. But the latter would die in the arid regions to which the desert dog is adapted. Desert dogs are thin and lanky so that they can run fast and for long distances in these harsh regions where their quarry is often as swift as the gazelle. All these different animals are carnivores. So too are the seal and the sea lion, which have adapted to another distinct milieu: water.

In a certain part of the Shetland Islands, where the climate is especially harsh and there is little vegetation, a breed of cattle has been imported that is small but extremely hardy, with long hair for protection against the cold. Having forgotten that they are herbivores, these cattle feed on . . . fish!

Animals that live on islands become progressively smaller. In most locations the horse survived for 5 or 10 million years without getting any smaller, but on islands it evolved into the pony. The large size and long legs of the horse carry no particular advantage on a small island.

An animal's environment is made up of water, terrain, and vegetation, but also of the other species that live there. And if prairies are necessary for herbivores to subsist, are not the same herbivores there to ensure the survival of carnivores?

The earth is divided into glacial, desert, temperate, equatorial, and oceanic zones.

Starting at the North Pole, we find a glacial region where the fauna is highly specialized. Next comes a region of conifers (Finland, etc.), followed by temperate climates and then by deserts. Next comes a zone of tropical forests: We have arrived at the equator. Then there are oceanic islands, the home of birds and tortoises. Lastly there are the seas and oceans, the exclusive domain of fish and of mammals such as the whale and the dolphin.

These zones are themselves divided into different regions: maritime, savanna, mountainous, and continental. The rivers and lakes of each region support different genetic forms.

Depending upon whether they are found on the prairies or in the mountains, herbivores are divided into two main categories:

Perissodactyls, which have only one digit, include zebras and horses, and are capable of traveling very rapidly over flat land; and

Artiodactyls, which have an even number of digits, and include cows, goats, and chamois.

Perissodactyls cannot adapt to mountains; their single digit prevents them from negotiating this terrain. On the other hand, artiodactyls can get about very well on rock because their double hooves cling to uneven surfaces.

In dry, desert regions like the Sinai, great numbers of lizards are found whose skin is so hard that it prevents any evaporation of body moisture. Thus, their skin functions in the same manner as closed automobile motors that never need to have water added.

Our friend the frog is found in regions with ponds and marshes. Like all other amphibians, he is the result of a long adaptation to stagnant water.

Of course, we are excluding artificial environments for which man is responsible. Thus, for example, the saluki may be found shut up in a kitchen, and the sled dog can bask in the sun on a Mediterranean beach!

"Tropisms" were discovered at the end of the nineteenth century by the German scientist Jacques Loeb. Loeb committed certain errors then that are recognized today, but the broad outlines of his discovery remain valid.

Tropism may be defined as an automatic tendency that draws an animal toward something. Tropisms are inscribed in the genetic memory of the animal. In the evening, when the lights are turned on, moths immediately fly in through the open window. This attraction toward light constitutes a tropism. Similarly a tropism draws a herd of cattle toward fresh pastures.

In general terms, tropisms are designed to assure the survival of the species. In lower animals the movement is innate. Neither intelligence nor reason can modify it, although it is always possible to alter the external stimulus. If instead of turning on an electric light one lights a candle, the moths will burn their wings on it.

The story of sea tortoises is one of the most extraordinary in the animal kingdom.

When egg laying time arrives, the tortoise leaves the ocean and takes to solid ground. Set in motion by a mysterious instinct, thousands of them ponderously haul themselves toward the place where they must lay their eggs. The black line of march of these gigantic chelonians on the white sand is a sight that has been repeated for 220 million years, like a ritual procession of immutable rhythm.

The tortoise arrives at the appointed spot. She stops. Then, with all the force of which her feet are capable, she digs into the sand in order to make her nest. Next she deposits her eggs there, by the hundreds. The task lasts for hours. Finally, the eggs are covered with warm sand, which serves both as an artificial covering and as camouflage against predators.

Exhausted, the tortoise seems ready to die on the spot. She can survive for no more than several hours outside her natural element. At this critical moment, a tropism intervenes. Spurred on by an obscure force that orders her to get going, the tortoise "makes haste slowly" toward the sea.

Extracting her feet laboriously from the sand that holds them, she drags her hundreds of pounds of carapace under the hot sun. By a heart-rending effort, by indomitable will, she heads toward the distant ocean, which is sometimes several miles away. She never errs. And once she reaches the first wave, she revives—saved!

The story is more astonishing still. The eggs are hatched. Out of the broken shells tiny baby tortoises burst forth. They know nothing about life. No mother has taught them what the land and water are. What is more, birds have been circling in the sky for several days, patiently awaiting their birth.

Nevertheless all the baby tortoises rush off in the same direction! Like thousands of ink spots on a white sheet, they run toward the invisible sea. They never lose their way or slacken their pace toward their vital milieu, even to flee the attacks of birds of prey.

The baby tortoises know what they need. The same tropism that acted on their unthinking mothers programmed their own genetic memories in the same way.

In 1945 an atomic bomb was exploded on Bikini. Bikini and its islets had always been a favorite site for tortoises to lay their eggs. But the radioactivity from the bomb persisted, and it erased the "computer program" for the mothers' tropism.

When they had finished laying their eggs, the mothers left randomly in all directions, to sink into the sand, yield to fatigue, and die. And their babies met the same fate.

Of all the types of tropism, one of the most important is the "chemotropism." This makes the animal react to a chemical substance, for example, catnip. The mere odor of catnip puts a cat under a sexual spell. The odor that a female butterfly emits announces to the male that she is in the area, perhaps as far as three miles away. And it is a tropism that makes him join her.

Odors may be sexual signals in man as well as in animals. Modern deodorants, regardless of what the advertisements claim, would then be antisexual! Why does a male suddenly go off in pursuit of a female? A chemotropism must be at work, released by the scent that the female emits. It is very likely that this would not even take place consciously (we have seen that man's sense of smell is the weakest of all), but this behavior, if it does exist, could still be programmed genetically.

Sometimes a responsiveness to contact with another body is what draws two animals together. In cases of this type we refer to "thigmotropisms." (How could this have evolved originally anyway, as a mutation in a single individual?)

A "thermotropism" is what impels the fish of the Moroccan coast to make their way to the Gulf Stream, the warm waters of which are conducive to their reproductive life.

When an animal is disoriented in a thunderstorm, we refer to a "galvanotropism," which is determined by the electrical field.

All of behavior, including that of man, is thus subject to tropisms. These are encoded in the genetic inheritance.

The reflex is an adaptive phenomenon that varies across species and individuals. A reflex can be conditioned ("Pavlovian reflex").

A dog that has had stones thrown at him will flee when

someone piles them up in front of him. But if the person has been rolling the stones toward the dog in play, the same act of piling them up will make the dog bark and jump with joy.

Even though it is a distinct phenomenon, the reflex is part of the genetic program. It is almost always modifiable in man and in higher animals, as though the engram has blank spaces that are to be filled in according to the animal's life experiences. The genetic program, however, remains irreversible, by definition.

There is another, less well known means of transmitting information to the next generation: acquired memories. This information is not genetically programmed, and yet it can be passed on for several generations. Nevertheless its transmission is time-limited.

Some remarkable results have been obtained in this area by Pavlov and by the biopsychologist MacDougall in his research on rats. After being subjected to various procedures, the first rats had acquired certain habits. Their descendants then received the same training procedures, with the result that the number of errors they committed decreased progressively with succeeding generations. The first generation had made 144 errors; by the 43d, there were no more than 20!

This experiment indicates that acquired behaviors can be transmitted from father to son, at least in rats. But rats, as we have said, are highly intelligent.

The same thing happens with higher animals. It has been noticed that animals born and raised in circuses perform much better than do recently captured ones, as though certain behavior patterns that have been repeated over and over are passed on to the offspring.

Nevertheless we are not dealing here with genetic information because, if the training is suddenly stopped, these animals will gradually forget everything that their ancestors have learned.

Therefore this form of information transmission is limited.

We have no idea how it works. It may very well be, at least in part, that the parents are somehow educating their young. A poodle, by observing his parents walk on two legs, may be impelled to imitate them.

The distinction between inherited and acquired behavior is difficult to make. In both cases, we are somewhat ignorant of the mechanism of transmission. We can only recognize that transmission does take place.

The field of genetics has allowed us to understand more about "instincts," a notion that we have used for several centuries to "explain" whatever remained inexplicable in a wild animal, a dog, or a bee.

Aristophanes defined instinct, but nothing was really explained by saying that the behavior arose from the essence of the animal. Thus, the biologist Grassé declared: "Instinct is the innate capacity to perform certain specific acts perfectly without previous learning (in bees as well as in babies), under various external and physiological conditions."

It appears that about 50 percent of the tropisms and reflexes that are encoded genetically fit the definition of instincts. This is a scientific fact. But other factors can intervene, such as neural and hormonal events. For example, the release of adrenalin, or epinephrine, in a feline causes the claws to become unsheathed; this occurs before any aggressive behavior. Doubtless many other such mechanisms exist; we only understand a few of them.

It is likely that an animal's instincts are closely related to its senses. The senses of animals, as we have seen, are much more developed than those of man. Animals receive information from the outside world and can interpret these messages immediately. This why we say that "instinct" permits a hunting dog to track its quarry. The dog relies as much on its genes as on its remarkable nose, which allows it to sort out odors and to identify the one that interests it.

An animal's instincts can be partially suppressed, and just as easily magnified. Fifty years ago, dogs, cats, chickens, and ducks did not comprehend the danger of automobiles, and so they perished in great numbers. Nowadays, if no more dogs die in auto accidents than men, it is because dogs automatically obey their instinct to avoid cars. This is even truer of country dogs, accustomed to going out by themselves, than of city dogs, whose behavior is modified by their being restrained on a leash, which causes them to rely on their master before consulting their instincts.

As for chickens, they avoid autos like the plague, from the time they are hatched. This is another example of the modification of instincts by external conditions.

Basic instinctive behavior (for example, parental behavior) is constantly being enriched by new experiments by nature. If an animal's behavior is altered and this modification proves beneficial, the change will be passed on to future generations. On the other hand, a harmful mutation will be selected against. This process, natural selection, has been occurring for millennia.

One of the most baffling and astonishing phenomena of nature is the organization of a beehive. Bees have spent perhaps a million years, maybe more, in perfecting the hive. In addition, many more mutations were necessary to refine fabrication of the hive by each generation. The bees' complex, cooperative behavior in constructing the hive is a tribute to the power of natural selection to shape behavior.

FOUR
The House and the City

You are a nobleman. You have your own country manor, with a park, on your estate.

Another nobleman, the lion, has his own den, where his lionesses and cubs live. He also possesses a park of 7,500 acres—his "territory"—which is off limits to every other male of his species. This "territory" is located within his "home range": the area that he knows, inhabits, and hunts, but does not defend. The lion is not unduly chauvinistic.

A modest little worker, the lizard contents himself with a suburban garden of 40 square yards.

The swan, with all the arrogance of the upper crust, generously allots himself a property of 295 to 370 acres—with a riverfront!

Except for species whose existence is communal, almost all animals, down to the tiny cricket, have individual territories. Here they live and forage, alone or with their families.

The concept of territory can be extremely complex. Territories may be overlapping, with "rights of way" at certain times of the day and along specified corridors that become the equivalent of common territories open to *all* animals, but only temporarily. Thus, in Kenya, the guide will ask you which kind of animal you wish to see drink, and then he will specify the exact time at which you should be at the water hole to see that particular animal. Each family or social group adheres to a regimen acknowledged by all and comes to drink at a different hour. One family of lions arrives at 7 P.M., another at 4 A.M., a third at 6 A.M. . . .

It would never occur to a lion to come to drink with the gazelles. Similarly, none of the latter would go to the river when a predator was slaking his thirst there. Everything is delicately timed—which is proof that animals have a precise notion of time and take it into account.

For a whole month, one owl showed its appreciation of time by coming, every evening, to converse with a person. The owl and the lady exchanged "hoos"—which seemed to fascinate the nocturnal bird. He arrived at 6 P.M. sharp and left 15 minutes later.

The most complex territories are usually those of felines. It is easy to appreciate this if one has cats, especially if they are not castrated.

A black Persian lived for eight years in the same house and the same garden. One day, to his horror, he saw a rival move in—and what a rival: a European tiger stripe!

There was an exchange of insults—very coarse language! Unsheathed claws raked the air . . . but at a safe distance: The felines were quite content with play-acting their aggression.

However, the jets of urine that Blacky—who ordinarily was very proper—sprayed on the walls were definitely not decorous. Nor were those of Tiger, who was eager to match Blacky squirt for squirt. Their respective masters spanked them repeatedly to "teach them some manners." But this accomplished nothing; moreover, it was unfair: A feline marks his territory so that other animals will know the limits of his claim.

A month later, all was in order. The two cats no longer ran into each other. The invisible barrier of their "frontiers" separated them. Even if they happened to pass within four inches of each other, they did not "notice" each other!

They had delineated their territories in such detail that the craftiest Norman peasant would have been completely confused. According to their agreement, the top of the buffet, which formerly had been Blacky's favorite spot, became Tiger's property. The large wooden table had been divided in two, and each of them placidly performed his toilette at one end—always the same end—without disturbing the other. But if Tiger set foot in the room that Blacky had always occupied, a fight ensued, characterized by insults, profanity, and orderly but hasty retreat.

Temporary territories also existed, such as the windowsill that was occupied first by one of them and then by the other. Somehow they never seemed to run into each other! Nothing is left to chance with felines. The temporary resting place that the window offered was utilized by the two cats at different times according to a tacitly established schedule.

Some time later a German shepherd invaded the carefully partitioned house, and some skirmishes took place, due to the language barrier. Then things calmed down. The dog was able to stroll about in peace; the cats saw nothing objectionable in this.

In much the same way, the territory of a tiger or leopard is inhabited by buffalo, elephants, birds, rabbits—all the animals of creation except for another tiger or leopard!

In Kenya a family of lions lived among some rocks. About a mile away, within the lions' territory, about 20 giraffes had set up housekeeping near some other rocks. The two communities paid absolutely no attention to each other and lived in peace.

A herd of elephants, crossing a path from right to left, passed within several yards of a buffalo that was taking the same route from left to right. By a sort of convention, or rule of etiquette, they pretended not to see each other.

The immense eucalyptus trees that grow in Africa provide an example of vertically-arranged foraging territories. These trees offer space and food to a wide sampling of animal species. At the summit are birds. Halfway down are monkeys browsing. Then come giraffes, next gazelles, and, way at the bottom, the wart-hog—not to mention the insects!

But why do two animals or groups of animals of the same species never live together? The answer, of course, is sexual jealousy: "Thou shalt not covet thy neighbor's wife." Doubtless there also exists in animals a sentiment similar to that of humans: A man living on his property has no objection to dogs, cats, birds, or ants spending time with his family. But when he meets an intruder, he cries "Thief!" If need be, he will reach for his rifle.

Possessing a sense of property like a capitalist, an animal has no more urgent need than that of marking the boundaries of his property as rapidly as possible in order to forbid encroachment by his conspecifics. To accomplish this, dogs and cats take advantage of their species' sensitivity to odors. They mark their territory with urine or with an odoriferous liquid produced by their anal glands.

In spite of being domesticated, the dog—perhaps even more than the cat—has retained his pronounced sense of territoriality (which makes him very useful to man).

As we have learned, Blacky and Tiger's pal is the three-year-old German shepherd, six feet tall when he stands on his hind legs. With his fangs, claws, bristling fur, lolling tongue, and demonic look, he terrorizes anyone who approaches the grillwork enclosing his territory *(his* house, *his* garden).

One day he discovered that, by seizing the handle of the gate in the crook of his paw and turning it, he could open the gate. Calm and self-possessed as a man of property, he escaped and, strolling through the streets, went for a little walk.

Everyone in the neighborhood ran for cover, trembling with fear. Only one foolhardy man remained out of doors to face the wild beast coming toward him ... wagging its tail, wanting to make friends! The fact is, once outside his territory and not needing to defend it any longer, a dog like this one is the gentlest, kindest animal in the world.

A male dog lifts his leg as frequently as he does when he is let out only in order to mark his temporary territory: the street. The next dog to come along has the same objective and immediately urinates on all the spots where his predecessor has, thereby replacing the latter's scent marks with his own.

The latest discovery of the modern canine is the territory on wheels. He does not allow any stranger to approach "his" car!

The smaller his territory, the more the animal becomes aggressive. This explains why, if a dog is kept chained up, he becomes vicious.

In the evening, in small towns, it is unusual not to hear a dog barking. In the distance, another answers him. Then a third. The evening concert reaches a crescendo and then gradually subsides. The dogs go to sleep. Each has warned the others that he is at home: It is useless to go out. Barking is a threat that precedes combat—or substitutes for it.

The big brown bear—whose sense of smell is doubtless less sensitive than the dog's, but whose eyesight is better—claws the trees that delimit his hunting ground.

In order to mark their territory at rutting time, stags scratch the trees with their antlers and deposit a secretion produced by glands next to their eyes.

The European bison (at least what is left of him, in a preserve straddling the Russo-Polish border) is a complicated case. Having worked on the trees with his horn, he urinates on the ground and then rolls around in the resulting mud. Then he rubs his body against the marks that he has made on the bark, plastering it with this special concoction!

As for domestic herds, for centuries their territories have been delineated by the dogs that guarded them. Thus it is the dog that points out their borders to them—by circling around them.

Wild herds of wildebeests, gazelles, and zebras resemble

nomads. Their territory and their home range (the zone where they graze) shift. When there is nothing more to eat, the herd sets itself up elsewhere, for a duration determined by the richness of the pasturage.

Communal territories are known also. It is strange to find living in such territories animals to whom the whole ocean is available: the pinnipeds and cetaceans. Dolphins, seals, and porpoises live in reduced spaces where they squabble like neighbors who don't get along. Petty gossip is followed by reconciliations and play. Sometimes a passionate drama suddenly is enacted: A baby has just died and its mother tries to steal someone else's baby. This is a pattern well known to psychiatrists.

Within these kinds of communal territories certain sea birds live also, such as the "Bashan fools" whose nests, at the time of egg laying, are so close together that they touch on the rock that serves as their collective residence. Each morning the males leave for work—they are fishermen—while the females remain motionless on their eggs. When they return, the males find the females just as they left them: quarreling incessantly. The females screech insults at each other and quiet down only in order to open their beaks for some sardines that their poor husbands hasten to throw in.

Actually, living in such close quarters aids the birds. If a predator that was fond of eggs were to venture into the village, all these shrews would cease quarreling and throw themselves against him—like "fools"!

It is hard to imagine that fish can maintain a territory in undulating water. Yet the stickleback—a little fish beloved by ethologists, who have studied him exhaustively—has succeeded in doing so. Once he has constructed his nest and has attracted a female to it, he stands guard and prevents intrusion by other males.

As for the famous Capitoline geese—they seem to think that this Roman hill belongs to them.

A territory is both a hunting ground and a place for strolling or resting.

It is a shade tree where the panther, dappled by light and shadow, is no more than a leaf among leaves.

It is a baobab against which the gigantic elephant sharpens his defenses, as a little cat sharpens its claws against the living room armchair. But eventually the enormous baobab, hacked to death by the deadly tusks, collapses, killing the elephant.

It is a berry bush where the fox, his white muzzle smeared with red fruit, gives full rein to his gastronomy and then stretches out in the grass to sun himself and take a nap.

These places can be retreats for avoiding predators. Alternatively they are favorite spots that are linked by "roads" that the animal has cleared. It is the strangest thing that a new inhabitant will always adopt his predecessor's pathways for his own. Thus the "roads" traced by large game are eventually expropriated by man.

Rare is the animal that does not possess his favorite spot: a tree, rock, cave, bush, warren, or burrow. His home is always situated in the best place; it is his castle keep.

The fox—a canine with the eyes of a cat—digs his burrow when he is about to take a mate, in the spring. The den will serve as a nest—since the mother will have her pups there—a resting place, a pantry, and especially as a strong castle for escaping from and defending against enemies.

Thus the fox's home reflects his strategic needs. This is especially true of its location. For days the fox searches for the ideal place: comfortable, secure, and not requiring excessive maintenance. The soil must be friable enough to allow easy excavation, and must have a firm foundation: rocks or roots that will prevent flooding and cave-ins. Moreover, the apartment must have good light! In short, the fox needs a spot that is dry and is hidden by a bush or branch, and where he can have a nice view without having to go outside and without being seen. It often happens that a den is abandoned in the course of construction. In such a case, something has displeased the architect.

At last, the fox has found the perfect location on which to build his three-room apartment! Excavation begins. He scoops out the dirt and piles it up to use for building up the entrance. Frequently a false den is dug far away to fool dogs. Underground, "communication trenches" stretch out for several yards, ending at the main exit. Four or five escape routes are concealed, against the necessity of escaping from the enemy. In the center of the labyrinth is found the living quarters, with vestibule, storehouse, and "castle keep": the matrimonial chamber that the vixen leaves, reluctantly, only after she has had her young, and that the male keeps filled with food.

The fox's den is the masterpiece of animal architecture. But it is hard to maintain; it is often full of rodents.

Affecting a different style, the beaver does not manage badly either. With his large tail that he uses as a paddle, he builds a dam. At the center of the artificial lake thus created he erects his domicile. There he will be able to while away peaceful days, safe from unappreciated visits from the wolf, the lynx, or the fur thief: man.

Some animals prefer communal living over solitude, or family living. Communal living does not imply that the individual's territory (or, as is common, the couple's) is an "open house" to other citizens. Rather, each individual or pair is ensconced in its own apartment just like in a city.

A society, whether it be human or animal, demands certain behavior of the individual, even if he retains his independence. The behavior that is expected varies with the group and the species.

It may consist simply of the animals' staying together, in species (voles, prairie dogs, wild rabbits) content to arrange their burrows into towns. These societies are actually anarchic, with neither leader nor laws, where each individual does what he pleases.

Other animals choose to come together only at a designated time of day. Who has not seen a tree become covered with birds at dusk, where there has not been a single one during the day? The winged throng cheeps and squabbles, discussing who knows what in the foliage. But gradually the noise subsides, and only an occasional cheep from a laggard or an insomniac is heard. Finally there is silence. Everyone is asleep.

That tree resembles a dormitory. Each bird there has his assigned place. He "crashes" there in the evening and leaves in the morning, flying off to his affairs without paying his neighbors any attention.

In general terms, three types of gathering may be recognized: seasonal, winter, and foraging.

To seasonal gatherings come animals driven by the need for foods that vary with the seasons.

In winter, many animals hibernate. For example, groups of bats assemble in caves. They hang upside down with their heads under their wing like skewered vampires. Likewise, thousands of ladybugs cluster in a field out in the countryside, where they pass the winter in a comfortable "dormitory."

Foraging animals assemble. These gatherings consist of either domestic animals that are never found together in nature—cattle, sheep, goats—or wild herds that will look for grass. Following inborn migratory routes, these wild herds sometimes make fatal mistakes. For example, some herds of wildebeests and elephants left Tanzania for Sahel where, because of the aridity, they died of hunger.

Birds and insects migrate through the sky. The migrations of swans, geese, swallows, and ducks are peaceful and predictable. A mysterious tropism causes them to cross oceans and continents twice a year as they go from their winter quarters to their summer retreats.

Other gatherings, annoying and even destructive, are unpredictable. These are the migrations of locusts, termites, and certain ants which for some unknown reason suddenly all leave the place where they are living.

None of these aggregations can be compared with human society. In order to merit the term "animal society," there must be a true need to live with conspecifics, an appetite for sociality different from but as powerful as the sex drive.

Organized, hierarchical societies have rules and laws that must not be transgressed under pain of punishment. Consisting of a variable number of animals, these societies provide a coherent unity where each individual has his place and must remain. The society occupies a communal territory and home range, and assigns a task to each member for which he is responsible—as we shall see in connection with wolves.

Animal societies are either monogamous or polygamous. In vertebrates, with very few exceptions, the leader is always a male. As important as the female may be (in elephants, for example, a female leads the herd of *females*), she yields to the uncontested authority of the leader. While the origin of male dominance is a subject of controversy in man, it is not disputed by animals. Female animals call to mind what Engels said of the female: that she was the first property of the male. And if a male animal forbids his "territory" to other males, it is largely in order to forbid to them possession of his female(s).

The only exceptions to this pattern of male dominance are certain communal, almost religious, insect societies (termites, ants, bees) that are led by the "queen."

The sexual significance of territoriality, and the aggression that it engenders, explains why a male does not select and mark his territory until puberty. Here we encounter testosterone, the male sex hormone that is responsible for many aspects of behavior.

The best example of animal habits and regulations—one might almost say of politics—is provided by the wolf pack. This animal society is organized very efficiently and intelligently, along military lines. (A less well-organized version is found in the form of the dog pack.)

The wolf pack members actually constitute a family and stay together for years. Females and young live together peacefully. Even the presence of strange males is tolerated. Everything is hierarchically organized. Both sexes obey the leader, who acquires this privilege by besting the other males in single combat. When defeated, a wolf lies down on his back in an exposed position. Actually this gesture is a display by which the vanquished acknowledges his subordination to the victor. Henceforward the subordinate will obey his superior in all matters.

Below the general of this army stands a colonel—the next strongest animal—and below him is his subordinate, etc. At the bottom are the weakest animals: miserable souls, buck privates, slaves—the doormats of the pack.

Within this patriarchal society, the she-wolves are led by the pack leader's female. But during the mating season the females isolate themselves with their male. The females take to a den to whelp. This may be a hollow in a rock, a bush, or some other protected spot. During this time, family life takes precedence over the social life of the pack.

The dictatorship of the dominant male is often challenged by younger wolves that, when they become adults, try to take his place. But the battles are often one-sided, and the "young Turks" usually find themselves quickly defeated. This goes on up to the time when the leader has become too old and must yield his place.

A dominance hierarchy has profound effects. It determines social relationships, neutralizes much aggression, and resolves conflicts that would otherwise disrupt the life of the community. The hierarchical order is stable; each animal obtains his rank through a defeat by or a victory over another. He can advance in rank only by defeating one of his superiors.

Even domestic chickens and roosters obey this law. The dominant rooster feeds first, walks ahead of the others—and has his pick of the females. The low-ranking hens go to the low-ranking roosters. It sometimes happens that a male undergoes "social castration." He is deprived of females and ends up by becoming impotent.

Wolves hunt like dogs (or is it that dogs hunt like wolves?): They have beaters. Their strategems astounded Hamilton, an English general of the last century:

> ... we saw a herd of antelope, and we noticed two animals that had crossed the plain. Wolves. When they were about 500 yards from the antelopes, they sat down tranquilly. After ten minutes the smaller got up and trotted off toward the hills. He appeared suddenly on the summit, running to and fro like a Scottish mastiff! The larger wolf, as soon as he saw that the antelopes were completely occupied with watching his comrade, got up and galloped toward the herd at full speed. It was obvious that the wolves made a habit of attacking in combination. The mission of the first wolf was to hold the antelopes' attention, and that of the other was to get to his feet stealthily and to rush into their midst.

Hamilton related that one day "... a wolf loomed into view, then a second, and then another. Fourteen of them set off, strung out in a line into the valley, like bobbers on the Seine ..." and cut down the encircled antelopes at will. This military tactic earned the general's admiration.

Living in packs and hunting much like wolves, wild dogs as

well as dingoes probably resemble the prehistoric dog from which our domestic dog is descended. Their behavior explains that of "man's best friend" toward his master, who is in reality his "superior."

Monkeys are rarely solitary. They love company. Their societies are as well organized and patriarchal as those of wolves, but are more democratic.

These are the only mammals that copulate at any time of the year, like man. That is why their social and sexual behavior patterns are inextricably linked. Although most of them are polygamous, these unions are permanent and contribute to the stability of the society.

Furthermore, the hands of monkeys allow movement and sensitivity that are much more highly developed than those of other animals. These capacities enable them to adjust quickly to any novel situation.

The social unit of baboons is the family: the male, his female(s), and the young. Frequently, celibate males are attached to this unit, forming the eternal triangle, so common in farce.

The "patriarch" dominates the females and celibate males, and has first choice of food, but if one of the males steals his food, the incident is considered minor. However, if he detects another male seducing his female—and she accepts the interloper—he loses his dominant position.

Physical possession is so important that at the conclusion of a fight between two baboons, the vanquished signals his submission by allowing the victor to mount him ritually. This is true regardless of the sexes of the baboons. The mounting is sexual only in its evolutionary derivation; sexual arousal does not take place. Male-like mounting means dominance, and female-like presenting connotes submission.

By submitting to the dominant male, a baboon gains his

protection against external attack and is integrated into the family group, where he may live in peace on the one condition that he does not flirt (too openly) with the harem females. If he is surprised in the act of violating this commandment, the traitor immediately releases the lady in question and submits to the husband in order to appease him.

A young monkey indulging herself in some liberties with an adult and surprised by the legitimate mate will not hesitate to "present" her sexual parts to the latter.

It can also happen that a female deceives—cheats on—her male and yet remains his obedient companion in public.

A dominant male abandoned by his group in favor of another male will present himself to the new leader. In so doing, he becomes a sort of viceroy, in that way getting his group back.

The late naturalist Sir Solly Zuckerman, who studied baboon sociosexual life very closely, thought that these sexually related dominance and submission gestures and privileges were also employed in prehistory by our human ancestors.

Chimpanzees are apes but their society resembles that of baboons, who are cercopithecoid monkeys. But we must note at the outset that, although the chimpanzee family is extremely unified and the young are very respectful (at 17 or 18 years, the adult sons still show affection toward their elderly mother), the family is really composed only of the mother and her children. There is no father—or rather there are too many of them!

This is due to the fact that a female, when she is sexually receptive (a 10-day span midway between menstrual periods which occur every 35 days), gives herself without hesitation to all the males who please her, even if there are 20 or 30 of them. Research on paternity, therefore, proves very difficult to conduct! We can help the chimpanzee males to salvage their honor by noting that they adore children. Perhaps each persuades himself that he is the father.

Despite everything about this primate that seems so amoral

to us, an embryo of Western morality does exist: Sisters and brothers refrain from mating, and sons do not copulate with their mothers.

From the time of her confinement, a female chimp is devoted exclusively to her offspring. She has no more sexual congress with the males for 3 to 5 years (during which time menstruation continues to take place normally). The resulting paucity of receptive females explains in large part the sexual behavior of the males.

The chimpanzee dominance hierarchy is about the same as that of baboons: dominant male, other high-ranking males, low-ranking males, juvenile males, and females ranked among themselves, with the ranks of the young corresponding roughly to those of their mothers.

The young males reach puberty at nine years. However, this does not seem to convince their parents that they have reached social maturity. They do not admit the young to adult society until 15 or 16 years of age.

As in baboons, a chimpanzee asks for protection or pardon by presenting. Chimps, however, are content to acknowledge this request with a perfunctory tap on the supplicant's hindquarters.

Since 1961 Jane Goodall has been studying the chimpanzees of Lake Tanganyika in their natural habitat (see *In the Shadow of Man*, Dell Publishing Co., 1972). This book is full of stories and descriptions of remarkably human behavior.

Of all the individual apes that she describes, the personality of the remarkable old mother, Flo, is the most memorable. This dominant female and mistress was so appealing to the males that all of them spurned the young "chimpettes" for a chance to win her favors. As is often the case, Flo attached herself to one particular male—but continued to make herself available to all.

Flo arrived with an imposing escort of admirers. Normally a sexual swelling does not last more than ten days. ... But this time it lasted three weeks, during which time her ardor did not slacken for an instant. It was during this time that we noticed a special kind of relationship between Flo and one of her suitors. ... The male was called Rudolph. ... He was a high-ranking chimpanzee. ... He became Flo's faithful companion. He escorted her everywhere. ... He stopped when she stopped, he slept in the nest nearest her [usually only the infant slept with a female] and it was to Rudolph that Flo ran when she was afraid or felt sick. ... But he offered no objection whatever when other males copulated with Flo. ...

Next the author describes to us a typical episode in the social life of these large primates: the rise of Mike: "In 1963 Mike ranked at the bottom of the hierarchy of adult males. ..."

Whenever a scuffle took place, he always wound up at the bottom: he was the last to get to eat. Nevertheless he was destined to surpass his peers at a single stroke—not by brute force, but by intelligence.

To signal their superiority, these primates engage in threat displays that frighten everyone, large and small—even though these are seldom followed by actual fighting. They puff themselves up, erect their fur, and arm themselves with stones and large branches.

What did Mike do? He noticed some tin cans in the humans' camp. These made much more noise when banged together than all of the usual chimpanzee carryings on.

> Suddenly he headed calmly toward our tent and grabbed hold of the handle of an empty gasoline can. Then he picked up a second one and walked about on his hind legs. ... He began to rock back and forth. He swayed more and more vigorously, his fur bristling, and he emitted a series of hoots. ... All of a sudden he charged

toward the group of males while banging the cans together.... Mike charged again, this time directly at Goliath. The dominant male jumped out of his way with as much haste as had the others. Then Mike stopped and sat down....

... Rudolph was the first to approach Mike ... he gave some little hoots of submission, and bent down and put his lips to Mike's thigh. Then he began to groom him. Two other males did the same.... Mike's use of man-made objects doubtless indicated a superior intelligence.

Goliath's nerves—he had been uncontested master until then—could not stand up to Mike's technological breakthrough for very long. "He rushed up to him, crouched down at his side whimpering, and began to groom him...."

Then Mike graciously (for animals can be big-hearted) groomed his vanquished rival.

From then on Goliath accepted Mike's supremacy. Moreover, instead of being angry at Mike, Goliath actually became the best friend of his delighted rival!

Birds are lewd! Like man, birds are voyeurs. Who, seeing the modest canary shut in his cage, would have suspected it? He is a born voyeur: The ornithologist F. Fraser Darling, observing seagulls, noticed that they were excited by seeing other birds copulate.

Success in laying eggs and raising young depends largely on visual and auditory stimulation. This is evidenced by the fact that laying takes place sooner in large colonies than in those with fewer birds.

American cuckoos are way ahead of bands of hippies: they have communal nests and child-care centers. This leads to promiscuity, as in man. Let's not say any more....

The pigeon refuses to lay unless a male is watching her. But a mirror in which she can see herself will do just as well! This trick does not work on seagulls, which require no less than 20 onlookers in order to lay their eggs!

An exception to these liberal ways, jackdaws have individual nests—even though they live in a highly organized society. They chase away strange jackdaws during the breeding season. Strangers, if they want to join the colony, can apply for immigrant visas in the autumn or winter.

All aspects of communal life influence the individual—transform him, educate him, instruct him. The wolf, with his beaters and quasi-military strategy, acts much more subtly, not to say more intelligently, than the tiger that, relying on force alone, is content to pounce on his prey. For everything that can be observed in animal life, there is a human equivalent. A society is more in a position to make progress than is an isolated individual.

However, animal society reflects biological intelligence, which can be extraordinary.

Mice and rats habitually store up surplus food. If the population of their "city" grows disproportionately, they immediately decrease their birthrate, either by fighting among the males or by decreased sexual activity—or by biological intelligence: The females have smaller and less frequent litters.

This phenomenon was manifested during the Second World War. Since food was scarce, the rats had litters two times per year instead of six to eight times. And each litter had two pups instead of four or six! When the war ended and food became plentiful again, births resumed their normal frequency.

Man is intelligent. If only he would show the same wisdom. . . .

FIVE
Solitude or Family Life

The marvelous preserve of the Ngorongoro Crater in Tanzania is truly a paradise on earth: In the midst of all the animals that gather there, live men who are convinced that they inhabit the kingdom of the gods.

Africa governs itself and governs its animals. Since the time when a certain Masai village lived symbiotically with its animal environment, two years have passed. The village has been transported out of the crater by the Tanzanian administration. This did not happen very long ago, but it is significant.

The Masai herders no longer have any right to do anything except water their herds in the crater, and now two permanent, well-worn trails can be seen. One has been made by the herds in descending into the crater, and the other in leaving it, like two overcrowded roads. Within sight of the procession are sly hyenas, nervous zebras, and unconcerned lions. Nowadays a man on foot can do nothing in this sanctuary. The mechanized tourist is king.

Here also lives a family of lions known to all the guides of the country. The male is a nobleman about 20 years old: powerful, slothful and superb, displaying in the sunlight the enormous beard-like mane that suits him so well. Occasionally he can be seen to leave his domicile and stretch out on the road in the shade of some auto. Unfortunately, man—an animal to which he pays no more attention than to others—has an exasperating mania for horn honking. This noise hurts the lion's ears so badly that he winds up getting to his feet angrily, muttering something into his Negusian beard, and returning to his personal rock. There he lies down and yawns, fatigued from the whole ordeal. The king of beasts is a lazy monarch.

Shortly before dawn, the wildebeest herd that passes not far away disperses. When dinner time approaches for the king of beasts, it is not a good idea to be too close to him. At this moment his mate and his sister-in-law interrupt their napping or attention to the young. They roar an announcement: It is time to go shopping—for themselves, for the cubs, and for the master. He himself would never so much as extend one paw for a gazelle. It was enough to have done so when he was a bachelor; why take a wife if you are going to keep working? He defends her—that is his role—by roaring once in a while to let the neighboring lions know that he is on his territory and they must not approach. They don't want to do so anyway; it would require them to exert themselves.

Lithe, stealthy, dangerous, with velvet paws and murderous claws, the huntresses start out in search of meat for the table.

He is vaguely conscious of the cubs, his sons, but he is indifferent to them. One of his daughters is in heat for the first time. Regarding himself as a pharaoh, he would mate with her if the mother did not oppose it. Yesterday he had playfully and innocently approached his daughter, but the old shrew had jumped at his head and boxed his ears. In all of his 20 years he had never seen anything like it—even when, as a young husband, he had cheated on his wife, who was pregnant at the time, with an unattached female in the neighborhood. He was smarting from it still! Women show their abhorrence of incest and adultery with a quick and vigilant paw.

They abhor incest—when it concerns their daughter and their husband. But what about her son, an inexperienced, two-year-old tyke who wants to show her a little affection? After all, it is she who has taught him how to fight, and making love is also part of his apprenticeship. Only this time it is the father who opposes incest. Scolding his son, he teaches him, with bites and scratches, about the filial respect due one's mother, and then he throws him out of the house.

The sex life of the lion is reduced to an hour of love about every two or three years! A male never copulates with a pregnant female (only man does), and since she is too busy with her young to have any more, she still refuses him the following year. In this she is wiser than lots of other females! As for the lion . . . since his sister-in-law lives with them, he finally works out an acceptable family relationship. As long as she doesn't actually see anything, the first wife does not raise too much of a fuss at finding her unmarried sister pregnant.

One can hardly talk about love, in the human sense, between this couple, but doubtless some kind of possessive affection is implied in jealousy. One might say, to commit an anthropomorphism, that this king lives as a bourgeois. He has the same mentality and the same hypocritical ways. Moreover, he is the only one of the great cats to lead a family life. Actually, nothing is more peace-loving than these 600 pounds of lion!

Gazelle for dinner tonight! The lion has just spotted the first of the huntresses bringing one down with a graceful leap. The lion stirs, exerting himself for a change, and roars that he is hungry. A little later he is gorging himself on warm meat. The lioness snarls and tears off a piece that she takes home.

What she has just sneaked off the table of her royal husband is not for herself. The lioness will be satisfied with whatever remains after her master has had his fill. Nor is the morsel for the cubs; hers are quite large now, and hunt for their own food. And the young who are sleeping over yonder belong to her sister, who will feed them when she returns.

With the gazelle quarter between her teeth, the lioness approaches another female who is as big as she. The other lioness has stood up, and she begins to purr and to rub against the huntress clumsily, following her around, and begins to eat some of the meat. No other member of the family would dare to come near in order to take part in the feast, neither an infant nor the husband. It is obvious that the big female is dangerous at this moment; she insists that the other female be allowed to eat in peace.

The other female is her daughter. Four years old and therefore post-pubescent, she should have left the family and gotten married by now. In any case, she should hunt for her own food and not be fed by others. At her age, this is just never seen!

The little lioness had still been very young when she had been surprised by a hyena—the wickedest, cruellest, most contemptible of all the animals. True, a hyena usually eats carrion. But if he cannot find any, he often attacks a living animal. Unfortunately, he does not know how to kill efficiently the way the large carnivores do. He tears a limb off his prey for his dinner and then goes off, leaving his horribly wounded victim to spend hours dying under the slow circling of scavenging vultures.

The little lioness had lost a rear paw in one such encounter. How could she have survived, with no human being to help her? Well, as everyone living in the preserve was aware, her mother had taken complete responsibility for her. The mother fed her, protected her, and left her only to hunt.

In this unusual case, normal parental behavior exceeded the simple instinct for nursing, protecting, and teaching the young. More was involved than hormones or genes, however we may characterize it.

Furthermore, this behavior went against two well-established principles of behavior. The first is that the offspring, when they reach puberty, stop showing the least interest in their parents and begin to lead their own lives. The second rule is that improvement of a species requires the death of all those individuals who come into the world sick, malformed, or abnormal.

It was obvious that the lioness had been born perfectly normal. She must have been five or six months old when the hyena had crippled her. But since she managed to reach the age of four years without being selected out, something besides the maternal instinct must have been operating. But what could this have been, in a heartless world where the unhealthy are consigned to certain death? Could the mother have felt in some vague way that the accident would not have occurred if she had watched her child more closely? Who knows? Perhaps she did not realize the seriousness of the wound.

This was an exclusively maternal matter, of which the father was aware but in which he did not play any part. This is normal. Aside from certain exceptions (fox, primates), the father animal is little concerned with the young. At best, as with lions, he defends them and feeds them when they are too young for their mother to leave them to hunt. For this reason the story of the two lionesses is extraordinary and is known throughout Tanzania.

Despite his reputation, the lion is a peaceful and compassionate animal, the only large feline with an appreciation of family—and of society. He likes company because he likes to show off, parading around in the sun in front of everyone—including humans, from whom in a few years he will doubtless beg food!

If the king of beasts is monogamous but inclined to deceive his wife, the leopard is solitary. When the time comes he takes a mistress. Then he abandons his pregnant female abruptly—if she hasn't shown him the door first! In any case, without regrets he returns to his favorite tree where he lives alone.

Few animals, especially man, have encountered a leopard without being killed. His spotted clothing camouflages him in the undergrowth where he has chosen to dwell and where, half invisible, he attends to his affairs. One tree branch serves as his bed and another as his pantry where he hoists his prey to protect it from the hyena. He does not take it down until lunchtime the next day.

The leopard provides a good example of "celibate" behavior. In the breeding season the leopard joins a female, almost always the same one. They then give themselves over to amatory frolicking that is difficult to ignore. Ordinarily afraid of their own shadow, they act with complete abandon, to the point of becoming exhibitionists, totally oblivious to the dangers—such as hunters—that menace them for the two short weeks of their honeymoon.

For comparable fondness for solitude and celibacy, there is only ... the hamster! This tiny rodent goes to visit his wife only for a very specific reason. Immediately afterward he is shown the door. But this does not bother him at all. He returns to his burrow where he can inventory his provisions, like a miser. A male hamster never pays attention to his wife or his children. He does not even offer them some of the seven pounds of grain that he can easily set aside—as minuscule as he is.

The squirrel is a coxcomb. After seducing a young damsel and becoming a parent, he finds that he has married a shrew. He quickly builds a new nest in the trunk of a tree—but near his female's nest all the same. Driven out by his wife's scoldings, the old boy savors a nut by himself in the shade of his tail.

American cousins of the leopard, jaguars live as couples, and even occasionally in a ménage à trois, which contradicts every animal ethic: one must be either monogamous or polygamous! This behavior does not prevent the jaguar from being even more secretive—if this is possible—than the leopard. Arboreal, he feels comfortable only in large forests and seldom comes down to the ground, except to hunt and to drag his prey over to the branch that he has chosen for his home. Totally asocial, he fraternizes with no other animal.

The tiger is a semisolitary beast who vacillates between solitude and family life. He remains with his female until she has her kittens. He hunts for her and for them, and then divorces her when he decides that he is no longer needed. He takes another wife the next breeding season. Unfortunately, females are not plentiful enough any more to present him with much of a choice! On the other hand, the tiger boasts near perfection as a predator; he is self-sufficient.

The tiger is more social than the leopard. When he has killed a large prey, he thinks nothing of inviting two or three other tigers to join his feast. But let's not misunderstand this. The prey must be very large, and its odor must have attracted the other tigers from afar (since each territory covers 400 to 600 square yards).

Almost all animals with spotted or striped fur are more or less solitary and little inclined toward family life. Their markings seem to be for concealment and hence suit their taste for privacy. A domestic cat with a striped or spotted pattern will be relatively wild and solitary, as a general rule.

One form of family life is found in herds, but it is organized on two different levels. The females and their young obey a female leader, while the males are organized under one of their number, an uncontested leader of the herd. Almost all herds of herbivores and aquatic mammals (sea elephants, seals, sea lions) adopt this form of social organization.

When the breeding season arrives, sometimes pairs isolate themselves within the herd, as we have seen in wolves. But it often happens, especially in herbivores, that the male acquires a harem for the breeding season—an undertaking that is accompanied by rivalry and fighting. A stag easily has six or eight does, the Mongolian horse is satisfied with three or four females, and female sea lions assemble themselves into groups of nine for the pleasure of a single male!

This is the equivalent of human polygamy, where the men live by themselves apart from the females, who are grouped in a women's dormitory under their mother or the first wife. All the children remain there, with the females. But the daughters leave when they get married, and the sons when they reach puberty. The sons are then the exclusive charges of the father, who teaches them the male role.

In herbivores the social structure embraces the family structure, whereas in lions, among other species, the family structure constitutes the entire social structure.

As in the overwhelming majority of human societies, polyandry hardly exists in animals. No one can recall having seen, among wildebeests or elephants, a female with several males. Some primates, however, such as the chimpanzee, seem to be promiscuous. Might this not be analogous to modern "swinging"?

If there is an animal for whom family life is synonymous with life, it is certainly the red fox, one of the most intelligent, cunning, and gentle of creatures.

We have already noted his natural aptitude for architecture and his dexterity in constructing his home, which will shelter him and his family for long years. He rarely moves to a new home unless safety or human extermination efforts require him to do so.

He is an excellent father and no less an excellent—and faithful—husband, capable of defending his vixen and her young to the death. Every hunter can tell you a story about a red fox leading on a pack of hounds for hours, in order to draw them as far as possible from the den and his family. Sometimes he gives up his own skin in the process.

Thus, one male let himself be killed in an attempt to save his offspring. Alas, his sacrifice was in vain. The dogs, trained by man to become the enemies of their close relative, located his lair all the same. They flushed the vixen out. Would she choose a brutal death at the hands of man to being slaughtered by a fellow animal who had no excuse for being a predator? Suddenly the hunter saw her rush to the mouth of her den, as red and trembling as the autumn leaves bunched together there to conceal the entrance.

The man shot, at the same time realizing something peculiar: instead of facing forward like an animal on the defensive, her back was to him. But now she was prostrate, dead. The man, however, believed she was only wounded, for her head, he thought, was still moving. It was actually her pup, which she had held in her mouth, having saved him to the end from the death that she had confronted. Thus the hunter learned the reason for her peculiar exit.

If there is a heaven, animals deserve a place in it—at least as much as hunters—and the fox will be rewarded for his intelligence and courage.

The vixen is unequaled as a mother. When she senses that the time for giving birth has arrived, she stores up food for eight days and then shuts herself in the "keep" where she will deliver. Later, at the time for parturition, she will go out one more time for a meal that she will predigest before offering it to her nurslings.

The red fox is monogamous, as are other animals with close family ties. But he must first fight furiously with the other males before he wins his wife. This is the game of love for male animals. He marries for life, but every breeding season he has to fight other males who come too close to his wife if he wants to keep her. His fidelity, furthermore, is poorly rewarded: after her confinement the vixen—like many other female animals, including the dog—growls, shows her teeth, and forbids him access to the inner sanctum that she and her young occupy. She doesn't need him for food: she has already provided it!

Only when the young have opened their eyes at nine days will the father be allowed in to see them. He will do so with much admiration and with that quizzical look that all males exhibit when they first gaze upon their offspring.

Then he quickly goes off to hunt, and returns to his den. Hidden near the entrance and ready to detect the slightest danger, he keeps watch. Other animals are vigilant, but the red fox is so devoted and intelligent that he can perform great exploits of courage and deception. It is not for nothing that his adventures are prominently featured in medieval literature. Wolves and the majority of other canids exhibit the same behavior.

Mammalian nurslings serve their apprenticeships in life according to one of two absolutely opposite modes. And, depending on whether they are "nidicolous" or "nidifugous," their parents will also exhibit totally different behavior toward them.

A nidifugous animal literally "flees the nest." Threatened from birth by predators, he must be able to escape them right away. Examples are the young of the zebra, the wildebeest, and the deer—mostly herbivores, the game preferred by large carnivores. To avoid being eaten, the youngster can only rely on his frail legs, for the herd will flee and abandon him in the event of an attack.

Mother Nature, who is perhaps a better mother than the real one of a nidifugous animal, sees to it that the youngster is born with its eyes open. Twenty minutes later the little one is on its feet, staggering about unsteadily but bravely, and ready to follow mama.

Actually, almost the first year of his existence is spent inside his mother's warm body. For example, a mare's pregnancy lasts 11 months and 11 days, plus 1 day for every year of her age—there is never a mistake! All trainers say that: Horses are good at math.

Nevertheless, although not a bad mother, a female herbivore usually shows nowhere near the emotional attachment to her young that the mother of a nidicolous infant does.

Nidicolous species have much shorter gestation times than nidifugous ones: from one month in the case of the hamster to 120 days in the lion (with the exception of man and the gorilla, both approximately nine months). The newborn is thus much less completely formed than a nidifugous infant. In most species the young are born blind and deaf (although their eyes are open and they can perceive sound in utero).

And they are so tiny! Look at the kitten! And his cousin the baby tiger: as an adult he will reach 600 pounds but now he hardly weighs 3 pounds. The heart sinks to see him so completely helpless, needing so much aid and protection. "The heart" is his mother's! He depends entirely on her: for warmth, milk, and love. Like the vixen, she loves him enough to die for him. This attachment lasts for three years, the length of his education, just as in the lion.

Giving birth is the greatest event of a lioness's life. As for her customary passivity, what kicks are there in catching one's dinner? And what is there to do the rest of the time except sleep? But as soon as she becomes pregnant she is almost feverish with excitement; everything interests her and she must depart from her usual routine. Just as a woman knits an infant's garment, the lioness tenderly prepares the nest by tearing silky hair from her abdomen without a second thought. Physiologically and psychologically, pregnancy rejuvenates a lioness.

Once the infant arrives, he demands constant attention: nursing, cleaning, walks, and all those extra efforts that make him someone special. But of all these things the most interesting for the mother as well as the baby is ... play, which is so very important for mammals.

Later, as he gets bigger, he has to receive his education. He can no longer just daydream in the sun. The young animal is so filled with curiosity that, if he is allowed to, he will walk under an ostrich and be trampled, or find himself nose-to-nose with a porcupine.

The animal having the most extraordinary maternal behavior is, paradoxically, a herbivore: the elephant, whose young are nidifugous.

The elephant does not need a husband to organize her life. She can do so by herself. Moreover, there is a female leader. Outside of the breeding season the males keep to themselves and the females group themselves around a female that directs them. Elephants remind one a little of those English households where the ladies chatter away in the salon while the gentlemen empty a bottle of port in the study. But whenever there is a grave decision to make, everyone gathers around the old elephant who is head of the family.

The lead elephant will be at the head of a charge, followed by the other males, who surround the females and the young. He will select the next feeding ground and lead his tribe there. He too, being a sorcerer without equal, will point out the exact spot in the middle of the Sahel where water can be found. An elephant needs to drink about 50 quarts of water a day.

Although the male is totally disinterested in his progeny, his mate holds the world record for maternal devotion. Her involvement begins even before birth, since gestation in the elephant lasts 21 months—almost two years!

At birth an elephant calf weighs from 225 to 400 pounds, which is light compared with the five or six tons that he will reach in adulthood. Only the blue whale can boast of a larger baby: two tons an hour after birth.

The birth of an elephant takes place amid a very peculiar ceremony, the like of which is not encountered in any other species. As soon as the female begins labor, which will last for four hours, the females of the herd form a circle to keep the males out and so they can assist at the delivery.

The baby elephant comes into the world ... all pink, baby pink. At once the midwives pat him lightly in order to force him to breathe and bring him to life—just like the doctor's slapping of a newborn! As soon as he responds, the enormous, delicate, hand-like trunks help him onto his feet and then lead him to the maternal mammary glands. However, confused and annoyed by his own trunk which he doesn't know what to do with, the calf has trouble nursing. So his aunts have to teach him how to proceed.

They continue to aid the mother for the first few months.

Knowing that to do so would be fatal, they never let a calf drop from fatigue during the long trips to a new feeding ground. It is almost as though they lead him by the trunk, forcing him to make the next way station.

As soon as the calf is weaned, his mother must teach him how to drink. The first time, he clumsily plunges his whole head into the water, chokes, sneezes, and almost drowns ... without swallowing a single drop. So his mother sucks water into her trunk and squirts it into his mouth until he understands that this is the proper way for an elephant to satisfy his thirst.

Thus he receives a regular education, including getting his ears boxed occasionally when he is disobedient and tries to leave the maternal apron strings to visit a neighboring family of lions. And when he is three years old, his mother will teach him how to charge in order to defend himself.

There is nothing more entertaining—or more dangerous—than to come across an elephant conducting her youngster across a road in a preserve. She begins by trumpeting a warning. Then, like a policeman with his baton, she lifts her trunk high. Only after taking every precaution does she cross the road, her calf at her side. If a motorist were to refuse to yield the right of way, he would learn in a hurry what an elephant charge is like!

The social structure of monkeys does not prevent them from having a family life: sometimes polygamous (hamadryas baboons, howler monkeys) but often monogamous. Regardless of the form, distinctly "middle class" niceties are observed: Even if she deceives her mate, the female monkey continues to play the role of the upstanding and obedient wife, as we have seen.

These distant cousins of ours exhibit parental behavior that resembles ours more closely than does that of any other group of animals. Usually there is a single baby which is adored by its mother. In addition, unlike the case in most species, the males also pay attention to the infant, showing it affection that goes much further than simple protectiveness.

At the turn of the century, when baboon behavior was poorly understood, one of these monkeys astounded an explorer. His dogs had scattered a baboon troop. The older monkeys fled into the trees, but one inexperienced youngster sought refuge on a rock, which was immediately surrounded by the pack.

In the face of this danger, the little one began to wail and extended his arms toward the trees. Upon seeing him, an old baboon—doubtless his father—quickly descended to the ground. Urged on by the cries of the whole troop, he clapped his hands, grunted, and stamped his feet, doing everything within his power to frighten the dogs. And in fact the dogs were so disconcerted that it did not even occur to them to attack the heroic simian when he clambered onto the rock, seized the youngster and, employing the same tactic on his way back, rejoined the troop in the trees. Quite a demonstration of courage and paternal dedication! A human father would have been proud of such an act.

The most "human" of the primates are in fact the chimpanzees, perhaps our closest relatives. Whereas the infants of other primate species grab onto the mother's fur as best they can, the mother chimpanzee often carries her baby in her arms, until he is old enough to ride on her back. When he is five or six months old and begins to walk, she takes him by the hand—just like a human mother!

Penguins also constitute exceptions to the rule that males do not care for their offspring. In these animals paternal behavior reaches heights of self-denial that seem abnormal!

Just as crazily, the female penguin prefers to lay her eggs in the dead of winter—and in the Antarctic, which is even sillier! Then she leaves, doubtless feeling that she has done her duty.

Then the male—yes, the male—replaces her. He carries out

his assignment of incubating the eggs for 40 days, without moving a feather! Eating only what he can reach with his beak, namely snow, he survives this ordeal.

He expects his wife to return with provisions. This she does, approaching with a fish in her beak ... for the young hatchlings!

Later, the penguins will follow the much more practical system of sea birds: the crèche. One adult remains on land to watch the young while the parents go fishing.

The only animal to match (almost) the splendid penguin is the male ostrich. He broods the eggs during the day, and then his females replace him for the night. Since the females share the night shift, they obviously do less work individually than their lone husband.

There is, however, an explanation for this unequal distribution of duties. With his white plumes and black tail, the male is much handsomer but also more conspicuous than the gray females. Since they are almost imperceptible, the females can forage without much risk during the day and leave to the master the job of hunting at night in the darkness which conceals him from predators.

The behavior of this giant among birds is biologically based. A hormone is involved, prolactin, which almost certainly operates in penguins too.

The behavior of the male ostrich is actually more complicated. At the time of the rut, he is "invaded" by male hormone. It is due to this state of high virility that he throws himself headlong in pursuit of females. Male and female run one behind the other for miles, in a veritable ballet which Walt Disney depicted beautifully in *Fantasia*.

As soon as the male succeeds in covering one female, he moves on to a second. This passionate love-making lasts a month. Then the mystery develops, because the male hormone disappears! Another hormone replaces it: prolactin, which causes milk production in mammals and incubation in birds.

The male ostrich has even more prolactin than the female! The result is that it is he who builds the nest where the females that he has fertilized will lay the eggs. He will also assume the majority of the responsibility for brooding the eggs. And there are not just three or four of them, but 90! He "was" virile.

Actually, except for certain unusual mammals that we have described, the only good fathers are fish or birds.

Birds carry parental love to great extremes, and the male feeds the hatchlings as much as the female does. The feeding of "crop milk" by pigeons is a well-known example.

Interestingly, the more the sexes resemble each other, the more they share in familial duties. One notable case is the handsome great-crested heron, found only in Japan. The male and female look identical—which incidentally often results in the male's suffering some embarrassment when he mistakenly makes advances toward another male. . . .

The male heron begins the nest even before he takes a wife. He begins it but the female finishes it. Along the way, they will share in the responsibilities, taking turns incubating and then feeding the young.

The paternal care of fish is quite special, since it is conferred only before the eggs hatch. After that, the little fish will become big without the help of mommy or daddy.

This behavior is all the more astonishing in oviparous animals, in which the males receive little indication of their impending fatherhood. Copulation does not exist in these virtuous inhabitants of the depths. In most fish the female lays her eggs and then the male deposits his semen over them, the female thus retaining her virginity. Even so, he goes through a great deal of trouble!

The *Betta splendens* is a complicated little thing, as anyone with an aquarium knows. He begins by building his nuptial nest: He blows hundreds of air bubbles made out of mucus. Then, leading his female beneath him (how risqué!), he presses her belly amorously, thus stimulating the expulsion of 300 or 400 golden eggs, which he immediately fertilizes.

The *Betta* takes these eggs one by one into his mouth, with infinite care, and deposits each into a bubble. And whenever a little fry—who has barely hatched but is already anxious to begin his life—tries to escape, the *Betta* recaptures the little fellow and puts him back in his nest with marvelous patience.

Most other fish, it is true, do not exhibit as much sensitivity. The love that they show their progeny is expressed differently; they do not deign to amuse themselves with their young.

Certain ovoviviparous species are ogres. These species experience all the pleasures of love, for the male is provided with an intromittent organ, the gonopod, which is formed out of the anterior rays of the anal fin and with which he penetrates the female to fertilize the eggs. The female swells up with the growing eggs, and after several days the fry escape. However, they do not always escape the maternal palate!

Let us not generalize too far. Oviparous and ovoviviparous species often make excellent parents.

Some examples of mouth breeding are very interesting. From 60 to 200 eggs take their place in the mother's mouth. In this she shows laudable restraint, since she feeds mainly on the eggs of her conspecifics. Man cannot distinguish one fish egg from another. But the parents never make a mistake and swallow one of their own, even by accident. Since sea water is constantly flowing through their mouths, this implies a deliberate suppression of the urge to commit infanticide. Another remarkable thing is that their saliva produces antibiotic substances which isolate the eggs in a bacteria-free medium.

Presently something extraordinary happens. The egg no longer finds refuge in the maternal mouth, but the fry himself does—and for two months! And since it would be difficult to swallow a meal without swallowing a few fry at the same time, this means that the mother stops eating for two months!

Sometimes the father comes to the aid of his female. Then

these hundreds of fry are transferred from mouth to mouth so that each parent may eat. This too is an oddity: aren't these fish carnivorous ...? But they don't confuse their own offspring with others' any more than they did their eggs. Even when the fry leave the maternal mouth to go swimming, as in haplochromes, they can rush back in when danger threatens and not get swallowed.

The baby fish is less prudent. He easily mistakes a lure—providing that it moves—for his mother.

Let us save for last the most outstanding of model fathers: Alytes, known as the "midwife toad." The nickname of this noble amphibian derives from the fact that he massages his female's abdomen when she is about to lay her eggs.

Thus unburdened of her load, the female hops away gaily and has no more to do with her future progeny. But the male assumes responsibility for the 50 eggs that she has just presented him with. And that's not all. The midwife toad is a polygamous little devil. It is not uncommon for him to have three wives. He is a veritable maniac for eggs.

Therefore he is burdened with—and that is the right word for it—150 eggs. He carries them on his back legs in several five-foot strings, even though he himself hardly measures two inches.

All larded with eggs, the midwife toad continues leading his tranquil existence. He hunts and gives an occasional concert (he is a distinguished flutist), but he never neglects to take his evening dip, necessary for the development of his future children.

They will hatch during one of these rinsings. Ungrateful tadpoles, they will immediately flee into the water, without even thanking the father to whom they owe their lives twice over.

All of the participants in these ceremonies experience the appropriate reactions to this powerful visual stimulation, just as their human brethren do. As the zoologist Darling said, when these collective displays in seagulls, Bashan fools, or penguins are followed by a "solo" performance, its purpose is to maintain the group's sexual arousal. In many cases the display induces the maturation of the female's gonads: She is generally a little behind the male in this respect. It is quite certain that the sight of birds copulating incites the others to do likewise.

In the midst of this teeming sexuality, true love thrives, expressed by tender caresses with the beak, gifts of choice insects or grain, and passionate quarrels—for aggression is often a sexual stimulant. Such "sentimental fools" are the lovebirds, the little parrots that can live only with each other and for each other.

But "bourgeois" values can temper both eroticism and sentiment. Thus, the females of certain hierarchically organized bird species, hens for example, delight in making the "perfect marriage," from which they draw a degree of self-satisfaction that might be called (without meaning to be complimentary) human.

If a hen that previously has been an outcast is chosen by the dominant rooster—the cock of the walk—she will quickly become insufferably arrogant. She will chase the other hens around and take precedence over them in eating and roosting. Because of these benefits she will wind up more beautiful. Thus, unlike the woman who makes herself attractive "before," the bird grows more beautiful "after."

If, in man, coquetry is the lot of the female, in birds it is usually the male who reserves that prerogative. The handsome males may boast sumptuous plumage, to the charms of which they will eventually add those of the voice. Even the least endowed of them possess essential embellishments—egrets a spot of color, for example—by which the seducer is recognized.

You may have noticed how the peacock uses his superb, fan-shaped tail in courting the female, who herself, poor thing, has no decorative appeal. Similarly, the canary sings with the shrillest tremolos, while his mate contents herself with peeps that connote fear or aggression.

The "fighting quail" and a type of Arctic bird, the phalarope, are exceptions to the rule. The females are more beautiful than the males, which are left the task of incubating the eggs. The females, furthermore, are the defenders of the territory.

This sort of matriarchal behavior is very rare in the higher animals. With his Medici collar, which gives him something of the formal air of a Spanish grandee, the lion provides the best example of masculine supremacy. This pattern goes all the way back to the little cuttlefish, a very ancient creature which, during courtship, decks himself out in black and sepia stripes. Then, in order to show off this wedding costume to better advantage, he spreads out all his tentacles, thus performing a strange sort of dance like something out of a cosmic ballet.

However, the female cuttlefish proves to be polyandrous. Rebelling against male supremacy, she requires as many as three males to fertilize her eggs, without gaining any more pleasure in the process and even incurring jealousy.

Actually, penetration occurs only in certain ovoviviparous sea animals, such as the lesser spotted dogfish, who wraps himself around the female to prove his love.

Two families of monkeys, the drills and the mandrills, have the most unusual ornamentation. Nature has painted their sexual organs red, blue, and green, suggestive of the ritual decoration used by primitive tribes, and also of the make-up of the Mayans or Egyptians. This makes one wonder if artificial bodily decoration is the successor of natural coloration of prehominid genitals. After all, man is just a distant cousin of the monkeys.

Even today, the breasts and pubic region of brides are decorated on the day of their marriage in many African and nomadic tribes which, even though relatively modern, practice this custom. It is also seen frequently among streetwalkers. In the developed countries, does make-up not serve as a seductive attraction, a sexual lure?

Young women who do not wish to use make-up are doubtless showing their rebellion against sexual subjugation or even a rejection of eroticism itself which, in their eyes, has been vulgarized and hence demystified.

All forms of sexuality are possible in animals, from masturbation to group sex and including homosexuality and the simultaneous orgasms of hermaphrodites.

The tapeworm, one of the most primitive species of animal, fertilizes itself. Its posterior segments are female and its anterior segments are male. To reproduce itself, it curls up so that the first segments face the last.

The snail, that well-known hermaphrodite, has strange customs. If Zeus gave birth to his daughter Athena through his forehead, the snail releases its eggs and makes love . . . through its ear! The proud owner of two sex organs, it introduces what might be considered a very primitive penis into the female opening of its partner, who returns the same favor. This having been accomplished, each goes its own way to lay the eggs that the other has fertilized. Therefore, if you happen upon two snails lined up horn to horn, do not disturb this tender scene.

Another astounding hermaphrodite, the oyster, changes its sex according to the season.

At the opposite end of the phylogenetic scale, a six-ton-elephant makes his partner get down on her knees so he won't crush her. This union takes place far from the herd, which is left for this purpose alone: Elephants are very modest. The male does not even expose his testicles, which are undescended; a rare situation in mammals. His penis weighs about 90 pounds.

The mongoose is just as prudish. Well, almost: He does

expose his testicles at the moment of copulation—not to prove to his companion that he is really a male but to chill them before usage! His testicles must be two degrees lower than his body temperature.

The elephant's virtuousness is not confined to his clandestine trysts. An elephant is also forbidden to copulate before he comes of age: at 18 years! Furthermore, he will copulate with the same female only once every five years, since gestation lasts 21 months and the education of the infant about three years. Happily for the male, elephants do not insist on fidelity. And then he can always indulge in solitary practices by making use of his trunk. ... Masturbation is actually very widespread among the animals. The dolphin rubs himself against the carapace of sea tortoises, which don't seem to mind, and the macaque likes to caress her own breasts and what Diderot delicately called her "jewel."

The dog, accused of every sort of vice, is what his lifelong companion, man, has made him. Whether he is a mutt or a show dog, he may be deprived of female companionship throughout his life. Masturbation and homosexuality will be poor compensations, for which it is hypocritical for us to reproach him.

And what can't dogs think of! A charming male dachshund was smitten with a female boxer. Their difference in size was truly frustrating. To solve the problem, the female rolled over on her back and the male ... acted like a man!

Two male greyhounds lightly exchange favors. Cats who have been castrated when very young like to mount each other or to be mounted like females. These are trifling matters.

The goat and the ram are two satyrs on which the Bible heaps scorn, unlike the shepherds who take care of—and are taken care of by—them! Lean, gaunt, self-assured and oh, how smelly, the goat is the sultan of about 40 females. He is justly proud of his attributes: proportionately, it is he who has the lion's share! The she-goat—perhaps to curry the male's favor?—

generously gives 40 times her weight in milk, thereby surpassing the record of the best milk cow. This superproduction also earned her the honor of nourishing Zeus. Man too owes much to her: She was one of the first animals to be domesticated.

This curious animal, after nursing a god, became the companion of sorcerers in the Middle Ages and was often burned with them in the town square. She is frequently hermaphroditic, and she sometimes gives birth to two-headed kids!

The world of the capricorns is supersexual and very peculiar. However, the long gestation of the single young at a time prevents excessive multiplication.

Very virile, the donkey is so sure of his enormous penis that he is immune to jealousy. Often, while he is waiting for his master in some Arab market, he mounts a female and then, satisfied, gives up his place to a friend.

However, he can prove to be vicious. A farmer owned a superb donkey stallion, to whom he led a mule mare. The donkey turned his back on her in disdain: This female was no donkey, so he wasn't interested. They were left together, with no results. Later, quite by accident while the man was fixing some chains, the noise so excited the donkey that he rushed toward the mare and inseminated her on the spot.

This was not just a coincidence. Subsequently, all that was necessary to get the animal to do his duty was to rattle some chains. Let Pavlov and Freud try to figure that one out!

But Nature herself, in her wisdom, sometimes reveals meaningful perversities. A magnificent bull was so prized that it was decided to use him only for artificial inseminations. He was presented with a cow decoy, that is, an animal maintained in a state of constant receptivity "thanks" to hormones. At the right moment, this animal is replaced with an artificial vagina.

With his lips pulled back, the bull smelled the cow as usual for a long time ... and then turned away. But to lose this precious semen would be disastrous; it was worth more than its weight in gold! What did the bull want? He would be given it! It soon became apparent: He wanted another bull.

But no male wanted to play this game. Finally, one was immobilized by harness, so no resistance could be offered. In this story, who was the more "vicious," man or beast?

Fertilization requires perfect coordination among the various glands that provide the vehicle for the sperm. The glands must secrete a diluting fluid, the prostate must produce a liquid that activates the spermatozoa, and the testes must produce the spermatozoa.

The dog, among other animals, does not have a coordinated glandular apparatus. His glands "go off" one after the other. If he withdraws too quickly, the meeting of sperm and ovum does not take place. To correct this "manufacturing defect," two erectile bulbs wedge the penis and keep it in the female until the end.

In felines, on the other hand, everything is very well coordinated. But ovulation can only be triggered by a neural reaction. To obtain this effect, the cat's penis is shaped like a bottle brush; when withdrawn, its cornified papillae bristle up and cause a sort of orgasm in the female. Is it pleasant or unpleasant? Only she knows!

The primary purpose of sexuality is fertilization of the ovum. For this to occur, Nature not only has to show imagination, she must also create external or internal signals that induce the animals to find each other and go to it—often very quickly (the mayfly barely lives one hour).

Various means of synchronization by olfactory or auditory signals, absolutely undetectable by man, enable "lower" animals to come together. Since this important and recent discovery has been made, some puzzling behaviors seem to be explicable—but not all.

Thus, in 1922 the zoologist Mell performed the following experiment on butterflies. Placing a female in his garden, he captured some males that were attracted to her. Then he released the insects, with identifying marks, at distances up to seven miles away. All the males returned! It is believed that an olfactory attractant is involved here, but no one knows for certain.

The lepidopterist Labonnefon once captured a female great night emperor moth. Soon his office was so invaded by male moths that he had to shut the window. The moths then entered through the chimney!

The auditory signals of insects are no less astonishing. All ten thousand different species have their own love call. No one knows how they recognize it, but no grasshopper ever heads for a cricket.

This strange and mysterious insect world, which includes 70 percent of animal species, features internal fertilization without coitus. On the other hand, many insects die from love—many male insects, that is. Everyone knows about the tragic destiny of the husband of the praying mantis. Being slyer, the male spider avoids the same fate by offering his female some food, which she chomps on merrily during the copulation.

Here the female is queen. The queen bee, whose nuptial flight is familiar to all schoolboys, exhausts the resources of the drone in one shot. She is provided with a curious nongenital pouch, a sort of "reserve tank" where the chosen drone deposits sperm by means of a primitive penis. He deposits enough sperm for the "mother" (who is prolific but not above reproach) to lay fertilized eggs throughout the two or three years that she will live. As for the drone, he has nothing more to do once he has admirably fulfilled the single duty that these Amazons demand of him, so he dies of exhaustion—and no wonder!

The "nest of vipers" is no more than a serpentine variant of the honeymoon.

During the breeding season the females gather in clusters to attract the males. They mix together and cross each other's path, while they await the arrival of their male conspecifics. "The men" arrive and try to break apart the nest in order to get inside. The cleverest slips beneath some female. He entwines her with his tail and the two of them remain in this position for hours. If vipers are referred to as lascivious, it is because they are good lovers: They are patient.

Suddenly the male bites his mate violently. She reacts immediately by opening her cloaca. He then penetrates it with one of his penes ... for he has two of them! Copulation resumes; it will last for a very long time and will be repeated. Now there is something to make men envious!

But it is the frog who once again provides the connecting link between the lower and higher animals. The frog, and his cousin the toad, seizes his wife in his hands; he holds her tight, gripping her sides. Under the influence of these sweet caresses, the female releases her eggs, which the male covers with sperm as they appear. This is the initial development in the evolution of coitus.

Do you know how to recognize a male frog?—by his third finger, which is longer than the female's. This too is a sign of sexual differentiation. Now try to determine the sex of a snake!

The "higher" animals manifest their sexuality by internal developments (increase in prolactin, other hormonal changes, ovulation) and external events (emission of olfactory or auditory signals, seeking of a nesting place and a mate, alteration of appearance). These two levels are inversely related to each other. With the evolution of coitus, some of the poetic side of love seems to have disappeared: singing, dancing, ornamentation. Pageantry was replaced by steadfast pursuit by the male. On the other hand, a form of pleasure seems to have evolved, such as that which humans experience.

We have described "normal" sexual behavior, masturbation, and homosexuality. We might also mention oral-genital contact: Only mammals indulge in this practice. The white rat pursues the female, catches her between his paws, licks her ardently, and lets her go so that he can pursue her anew. This lasts for an appreciable length of time before copulation takes place.

Similar prenuptial rituals or games are practiced by the female squirrel and cat. They roll around, rubbing their vulvas on the ground so that the odor emanating from that region will excite the male. Cries and meows accompany this display, which can last a long time in felines. Another pattern is mock fights. The marmot, for example, likes very much to follow his long hibernation with fierce jousting.

In almost all mammals copulation is performed stomach against back, man being an exception. Regardless of what modern theories of sexuality claim, to the effect that man originally made love like any self-respecting quadruped, the normal, natural position for human reproduction is stomach to stomach.

Man shares this style with some apes, the sloth, and . . . the hamster! The only difference between man and the hamster is that this spirited rodent easily copulates, stomach to stomach, 70 times in a row: the existing endurance record.

Let's not even talk about another tiny rodent, the midday jird, who is capable of copulating every 30 seconds . . . for two hours.

SEVEN
Yes, They Can Speak

The place was Kenya, on a warm, dry, December day. Cars were stopped on the road, their drivers taking photos.

With a leap, a lioness had sought protection behind some of the cars. She seemed no more ferocious than a cat that had hidden behind a piece of furniture in order to ambush a mouse. In sum, she found these vehicles very practical for hiding behind, and they had become part of her world.

Her particular "mouse" was a young, independent-minded wildebeest who, having strayed away from the others imprudently, grazed in the bottom of a ravine. Jumping from car to car, the lioness approached her prey. However, another wildebeest on the crest of the ravine where the rest of the herd was located spotted the feline. Letting out a soft cry—a warning signal—he fled. At the same instant the lioness leaped and made a fool of herself: Her quarry had fled and was now too far away to catch.

The lioness tried to recover her dignity, but she was aware of how silly she appeared. She looked sillier still as she hit her sides with her tail and headed back toward her rock home to swallow her anger—instead of a wildebeest—in the warm sun.

A little farther away, and a little later, an elephant had come to drink at a watering hole. Suddenly, a sound behind him disturbed him and he stopped drinking. What could it be? From a dark thicket another enormous pachyderm appeared. The first animal touched the bottom of the other's foot and lifted his trunk again, satisfied. It seemed to be an elephant, all right. But just to make sure, he felt the crook of the other's leg before relaxing his defenses. This time there was no question: He was a friend. The two comrades proceeded to lift a glass together like boon companions.

Another sound. The buddies turn around. A rhinoceros who seems to have no sense of timing (or is just thirsty) comes up to the fountain, too. Two gray masses bar his way. He seems to hesitate, and so the trunks are hoisted up. All the animals of the jungle know this aggressive signal. It produces a speedy reaction. Respectfully the rhinoceros takes two steps backward, turns around and hustles away at a good pace. One must never disturb an elephant that is drinking ... let alone two of them!

The wildebeest's warning cry, the elephant's raised trunk: These vocalizations and gestures constitute animal language.

"Ah, doctor," sighs the elderly lady in the privacy of the veterinarian's office. "That dog lacks only language!"

Twitching the end of his tail, the fox terrier turns his ear to the right in order to show that he can talk just like Mother and Father. His mistress takes no notice of it: She understands dog language no better than she understands Spanish or German. But although she would readily admit not knowing Spanish and German, it would never occur to her that she is ignorant of this other language also—which nevertheless is as useful as the other two! Actually, in her naïve anthropocentrism, she would never imagine that her darling dog could speak.

Obviously he could not speak in the literal sense of the term, to "articulate words." But lexicographers themselves acknowledge that "speak" may be used loosely to mean "convey one's feelings in ways other than words." Does a dog not do this? And all the other animals?

The dachshund who wags his tail is saying clearly, "I am happy to see you." But if he growls, the most obtuse person can tell that the dog is insulting or threatening him. This same person will shrug his shoulders when someone tells him that dogs can communicate.

People often confuse spoken language with means of communication, and are wrong to do so. A "tongue" is only one language among many, including animal languages. The courteous lion roars before embarking on a hunt, as if to say, "Beware, Mr. Prey, your master is hungry!"

But while he speaks, the lion twitches his tail, in an eloquent but entirely gesticulatory expression of his aggressive state. So as not to exclude any such dialects, ethologists use the term "communication" to include all signals between animals.

The language barrier does not seem to be limited to man's Tower of Babel. The languages of animals have differences, too, which impede understanding. There is a French expression, "to understand each other like a dog and cat," which means to understand each other not at all. And, in fact, no two things could be more different than a dog and a cat. The latter is an introvert, very much preoccupied with himself, that likes dreaming, silence, solitude, and his home. The dog loves noise and activity; he is bored being inside and always wants to go out and run. He is an extrovert.

But these differences would still be comparatively minor if their languages were not precisely contradictory. When the dog wags his tail, the cat thinks him furious, since for cats this is a sign of anger. And when the cat begins to purr, the dog believes that he is growling (growling and purring have the same vocal origin). The two species seem to mimic each other: but creased eyes, retracted ears, muzzle forward, and mouth open expresses aggression in the cat and tenderness in the dog. And so it goes.

Imagine a stranger who, when you told him you liked him, would think that you hated him, and who, when you invited him to visit would think that you wanted to throw him out, and who, when you thanked him, thought you were insulting him. How can a dog and cat, speaking such different languages, understand each other? And yet, especially if they are young, both will come to support and even sympathize with each other in a short time. Just like a person who has learned a foreign language, each of them will remember the other's means of expression. These supposed archenemies are capable of becoming true friends.

Youka, the German shepherd mentioned earlier, had as a friend an enormous blue Persian, Yahn. Having been raised together, they were of the same age and understood each other perfectly. Being the more astute, the cat had quickly learned how to open a door. The dog watched him do it and admired his handiwork but would stare, like an idiot, at a door that refused to open.

One day, Yahn had had enough. He decided to explain to Youka what to do. Seated on her rump, the puppy watched Yahn intently. Slowly, gently the cat slipped his paw into the space between the two halves of the double door. He paused to make sure that his friend had gotten the idea. Then he pulled the door toward him, remained in that position for a moment, and finally swung the door open, looking at Youka all the while. Shortly thereafter, the puppy repeated exactly the movements that she had just learned from her friend. Doubtless very interested in whether or not his lesson had sunk in, the Persian watched what she was doing. It was obvious to the people who saw this that the cat had provided a gesticulatory explanation that the dog had understood very well.

Language, whether it be vocal or gesticulatory, plays an important role in animal societies. It promotes cohesion, survival, reproduction, and respect for the hierarchy and also the training and education of the young.

By "giving tongue," the pack passes the word from dog to dog concerning the game that they are chasing together. This communication is accomplished by using very different barks which convey both a problem and its solution: "The rabbit's path is in the middle (problem); I'll take the left side, you take the right (solution)." In this way the dogs keep each other informed.

The same thing occurs in groups of wolves, jackals, or coyotes. Human speech is not always capable of doing as well since, unlike animal communication, it adjusts problems of sociability more than of society. Animals, on the other hand, seldom speak if they have nothing to say—that is, except for animals that have been spoiled by man and wind up resembling him, such as dogs and cats that like to chat with their masters.

For example, a precious Siamese cat named Dove sat opposite her mistress and exchanged "meows" with her. Dove accompanied these sounds with ear movements and then tail movements, which became more and more nervous. By using words that the cat did not understand, was the woman insulting her? Did Dove think that she was being made fun of? Suddenly, with an angry clawing of the air, the Siamese put an end to her mistress' cackling and took off, visibly upset.

Tiberius, a Belgian sheepdog, waited every morning for his master to awaken. As soon as the man opened his eyes, the dog put his paws on the bed and rubbed his big head affectionately against his companion's. This was his way of saying, "Good morning." Soon this gesture came to be accompanied by a little happy groan. As the months passed, this cry grew louder and changed in form, becoming a kind of solemn song which lasted nearly a minute, just as long as the head rubbing did.

But this was "speech" in a domestic animal. In the wild, wolves, who maintain their individual prerogatives with an almost feudal sense of honor and pride, will not hesitate to thrash a vassal that dares drink before his superior. Each animal must be informed both of the rank and the intentions of his peers by the use of a very complex gesticulatory language, involving the ears, the tail, and a variety of facial expressions.

Birds are natural, unrepentant chatterers. But is their singing merely an expression of pleasure? Are they really talking? Is that cheeping in the birdhouse a recitation of the day's adventures? In any case, all that is necessary for them to fly away at once is for one of them to give a particular call. We do not know the meaning of the preceding sounds, but we are certain of the meaning of the last: It is a warning call.

On the same subject, the observations by Konrad Lorenz of geese have provided some important knowledge. To show its contentment, this bird emits a particular clucking sound composed of six or seven syllables. If this cry is shorter, it indicates an intention to fly away. When the entire gaggle makes the sound, all will fly off. Two sharp, loud "gang gang" syllables always mean that the goose is going to fly away in the next instant. A single "gang" of the same pitch causes the cackling to stop: It is a warning cry.

It is known that all migratory birds hold a secret council before departing, at which orders apparently are given and received. During the V-shaped flight of geese and ducks, when the leader is tired he employs a special call to indicate that he wants to be replaced. And indeed he is then replaced at once, discipline being rigorously upheld during these flights. This discipline has to be inculcated in the birds; and this can be accomplished in no other way than with signals that are equivalent to language!

Do birds that speak "human"—parrots, jays, lyrebirds—merely repeat sounds without understanding them? Or do they grasp their meanings? Konrad Lorenz believes that ravens and certain parrots know what they are saying (I believe this is also true for lyrebirds):

> Their pronunciation of human words also clearly reveals the gratuitous play which characterizes the singing of less highly evolved birds. But ravens' and parrots' own sounds have become astonishingly independent; it

is incontestable that they convey ideational associations that are fixed and almost [*almost!*] charged with meaning.

To bolster this quasi-certainty. let us recall a story told in *What to Do Till the Veterinarian Comes:* the story of Jacotte, the parrot.

My friend Jacotte belonged to a public official who came originally from Gabon. She was loquacious, tame, and something of a tease. She and her master had the same mischievousness and understood each other wonderfully well.

One day at the beginning of Autumn man, wife, children, and parrot were in the country. Jacotte was perched quietly on her master's finger when she felt a sudden urge to fly toward a low branch in a large tree. Once there, she saluted her family with a mocking "well, good-bye" and began to climb up the trunk.

Being accustomed to these "flights of fancy," no one got very worried about this. That is, until the parrot was called and adamantly refused to descend.

Night fell. Our public official placed the longest ladder he could find against the tree and bravely climbed after the bird. Jacotte seemed to be amused by all this. As her master climbed each rung, she hopped to a higher branch. So he had to give up. He thought that she would surely come down the next day, when she got hungry. But this didn't happen. For several days Jacotte remained in the garden and refused to come home; she chatted amiably with the family. Then, one morning she was not heard from.

Each member of the family thought to himself that a cat, or hunger, had settled accounts with the parrot.

However, she was heard from again. One evening a farmer burst in, very upset, saying, "There's a ghost in the cemetery yew! The tree has started talking!" The farmer was calmed down, and the official went with him to try to capture the "ghost." Alas, Jacotte had already taken off. Gloomy days passed, and the family was forced to admit that she had disappeared for good.

One day toward the end of the summer, a violent storm broke out. The man, his wife, and the children pressed their noses against the windows, supposedly to watch the branches bend under the assault of the wind and rain. Actually, each was thinking about Jacotte.

Suddenly, between two claps of thunder, a forlorn voice was heard crying, "Papa, I'm cold! Mama, I'm hungry, I'm hungry!" And a ball of matted feathers tumbled down in front of the door, whining in a high, piercing voice.

The door was quickly opened. Jacotte entered, glaring angrily at everyone, whom she doubtless thought responsible for her plight. Then, with one last "Mama, I'm hungry!" she hurried over to her feeding dish.

I did not make this story up or exaggerate any of it. It is authentic in all its details. It is obvious that the sounds "I am cold" corresponded to her state of being cold, and "I am hungry" to a real need for food.

We can understand a certain number of obvious signals. Among them, notably, are signs of recognition.

It seems that members of the crow family must go through five different steps before being recognized by a conspecific. They proceed with caution! The sequence can be described as follows. They sniff each other beak to beak, then the head (at the spot where the glands are located which produce an odoriferous liquid that is used to mark the territory), the nape of the neck, the genitals, and finally the shoulder. Then, having made sure they have not been fooled, they separate peacefully.

The horse bids welcome to his compatriots in an analogous manner, except that anal sniffing occurs in place of investigation of the head. The horse's pattern resembles that of dogs.

We think that we understand the language of dogs better than that of other animals, but ethologists have learned that dogs use 25 different vocalizations, without counting movements, to express their feelings. Some of these sounds are yet to be deciphered, and scientists emphasize that we are still far from understanding "dog language." Anyway, here is a glossary which will help you to better understand your favorite canine:

Bristling fur: A warning signal, an indication of anger and aggressiveness; the animal appears to increase its size in order to frighten its adversary.

Ears slack, tongue hiding the fangs: He is alert.

Ears forward, tongue pulled back, forehead wrinkled: He is very alert.

Ears hidden: Fear and insecurity.

Ears hidden in back, muzzle closed, forehead furrowed by concentric wrinkles: He is irritated.

Same as above, but muzzle open and teeth exposed: Readiness for combat.

Tail erect, head up, cringing walk: This is the bearing of a hunting dog and signals aggressiveness toward the quarry that he is sure of overcoming.

Tail erect and muffled growling: Announces the main attack.

Wagging tail: Good disposition.

Tail between the legs: Fear and subordination.

Licking: Mark of sympathy and affection (not seen in all dogs).

Pushing someone with the nose: Play and spontaneous allegiance.

Body stretched out, muzzle between the paws, rear end elevated: Want to play with me?

Lying on the back, abdomen up: Affectionate submission.

Bounding along, head extended forward, tail moving unconsciously: This is a joyful "hello."

Baying in the evening responsively with other dogs: Territorial patrol.

Howling incessantly: Desperation, expressed when the pack or team begins to be decimated, but perhaps also a love call if a bitch is in heat, or possibly a sign of derangement in certain cases.

Very sharp barking (that of a pack of hounds): Announcement that something has been spotted.

Barking of the same type, but louder: A show of force—"I am on my territory"—but without attack.

Violent barking without rage: A sign of anger.

Barking that becomes increasingly violent: The opponent is attacking or fleeing.

As for marking with urine, this is, like a signature, a quasi, almost official act by which the territory is delimited.

The language of the cat, that mysterious creature, is still less well understood. Moreover, it is much more complex than the dog's; in addition to their expressive movements, felines can produce 63 different sounds with which to convey their feelings.

Although incomplete, the following lexicon will allow one to converse with a family cat:

Yawn: A sign of satisfaction.

Extending the paws, with the body supple: "At peace with the world."

Whiskers and ears pointed forward: A sign of rapt attention (all sensory systems are mobilized).

Purring: Very great satisfaction and desire to be friends.

Rubbing his head and sides against you: Desire to seduce.

Arching of the back: Desirous of caresses.

Arching of the back with a peculiar movement of the tail: Discontentment.

Bristling of the fur: Same meaning as in the dog.

Tail like a bottle brush: A sign of fear, as yet without aggressiveness.

Ears down, whiskers forward: Anger.
Ears down, whiskers back: Aggressiveness.
Tail stretched out, whiskers forward, ears pointed: Observation.
Tail at right angle, with only the upright part moving left and right: Prelude to combat.
Tail curled around the body: Repose.
Both the genital and anal orifices concealed: Modesty.
A peculiar, unique meowing: Mating call.
A low cooing in the throat: Copulation.
A half-meowing, very sharp cry of visceral pain: Post-copulatory cry.
Very tender meowing: Parturition.
Soft, repetitious little meowing: Kitten's call.

Reminiscent of ancient choirs where the soloist began a chant which was taken up by all the choristers, the croaking of the toad or his cousin the frog is also a language. The first croak uttered is like a rallying of the collective voice of the males, who will spend the night singing romantic serenades that the females will not be able to resist.

The beaver employs a language of architectural terms which enable him to give precise orders concerning the construction of dams and huts. But he is also irascible, and if contradicted will slap his tail violently on the ground to show his fury. This is exactly why the rabbit stamps the ground with his rear feet!

Although we have some knowledge of animal expression, we remain in total ignorance about many sounds and gestures. We must acknowledge that animals receive and transmit messages that we are unable to detect. This is a little embarrassing, for although we cannot understand a great deal of animal language, animals easily pick up various human words that concern and interest them.

Thus, a German shepherd is able to remember about a hundred words. Furthermore, he is remarkably observant and can recognize certain gestures without ever making a mistake.

Uépo used to take the train with his mistress. He installed himself in the compartment, stretched out comfortably, and went to sleep, thereby attracting admiration for his good behavior. Then, all of a sudden one day, he got up on his hind legs, barked, and rushed into the aisle, jostling everyone. His mistress, who was reading, had just taken off her glasses; this movement, or ritual, was a signal to him, as clear as if she had just said, "We're here." Incidentally, babies are susceptible to the same sort of conditioning.

Anyway, because he picked up certain gestures too quickly, Uépo was fooled about their meanings. One day, completely by chance, his mistress had taken off her sweater just before cleaning his ears—an operation that filled him with horror. Several evenings later, when the young woman made the same gesture to get undressed, she watched with stupefaction as Uépo immediately took off and hid.

Indian mahouts know that the elephant responds to 20 different commands. Similarly, trainers and tamers are familiar with the words and gestures that they must use in order to make a circus animal, a tiger, or a simple poodle understand them. Today we no longer doubt that certain human words correspond, in the animal's mind, to particular objects or actions.

In May, 1959, an institute for research on communication between species was founded in the United States. Included among its founders, and as yet unknown to the public, was Dr. Jean-Claude Lilly, whose study of dolphin language was soon to become famous.

Dr. Lilly had a predecessor, another American named Dr. Gardner, whom Jules Verne (who must have been very well informed) portrayed in his novel *The Big Forest*. Jean-Claude Lilly's predecessor was obsessed with the language of primates. In 1894 he decided to reverse the time-honored traditions of zoos and to put himself in a cage in a forest in Gabon. His idea was not for gorillas to take their children on Sundays to see "man," but to study their habits and language in safety. He intended to be imprisoned for three months, but he fled three days later, devoured by mosquitoes.

Nevertheless, the brevity of his stay did not prevent him from publishing a *Pocket Dictionary*. Doubtless, this was intended for use by an explorer lost in the bush who wants to ask directions of some orangutan, or to ask for something to drink or eat in monkey talk. For instance, where can I find a drink, "Cheny," or food, "Whouw."

Actually, all that we know for sure is that primates possess the widest variety of sounds that the human ear can perceive (the sounds emitted by dolphins are almost all ultrasonic). Of all the primates, the most remarkable is the howler monkey who, delighted to produce about 20 tonal modulations with his larynx, likes to make his voice heard as much as does a diva! But the chimpanzee surpasses him with 32 recognizable sounds. Moreover, we are not at all sure that this ape does not possess a spoken language of the human type.

Nowadays, in order to communicate with the chimpanzee, man utilizes an entirely different technique than that of 100 years ago. Sometimes blocks with letters on them are used, and sometimes sign language for the deaf. This latter technique, perhaps the most astonishing of all, has been employed for ten years in Oklahoma. The subject in the research is Washoe, far more knowledgeable than the charming Judy of *Daktari*.

Washoe's "parents," the American ethologists Allen and Beatrice Gardner, raised her in captivity and treated her as a human child. But to communicate with her, they taught her sign language, always confining themselves to its use when in her presence.

The results of this experiment are astounding. Washoe learned a very large number of words. But she also understood that a single word, when combined with others, can have a different meaning. "Open," for example, can be applied to a drawer, a door, a window, etc. Washoe has gone further still, grasping the notion of abstraction. Thus, presented with a fragrant flower, she made the sign for "flower" and then added spontaneously another sign that she knew: "smell of food."

Today, Washoe "teaches" other apes, and her best pupil, Sarah, already surpasses her in certain respects. Sarah has actually added writing to her "speech": the square that she picks up represents a banana, the triangle means an apple, etc. Moreover, she does not have to "see" in front of her what she is "writing"; that is, she is capable of representing an object outside of her presence.

Aside from the experiments conducted on dolphins, this is one of the most important breakthroughs in ethology in recent years.

As for Dr. Lilly, in addition to trying to understand the whistling, mostly ultrasonic, language of dolphins, he has sought to teach them ours. He states:

> I have distinctly heard words and expressions repeated in an extraordinary register and with an extremely rapid tempo: "Three, two, three" (after these digits had been pronounced), "T, R, P" (after these letters), and a number of other imitations, less clear but so faithful in their rendering of rhythm, intonation, and human timbre as to be fantastic.

Is this an indication of an imagination run wild?

Another American, Dwight Batteau, conceived of a machine that seems to be something out of a science fiction novel. It was intended to convert human words into audible sounds ... for dolphins! Unfortunately the inventor died too soon to complete his work and there are no present plans to resume it.

Similarly, some scientific experiments in Florida have been virtually abandoned. They were costing NASA, which was in charge of them, the price of a rocket!

What about fish, which people believe are mute? Yes, they

can "speak"! We know this because of sonar. But, as with dolphins, we cannot yet understand the meaning of their words. Their language is not only vocal, it is also transmitted by means of chemical signals concerned with reproduction and other behaviors. This is apparent from the fact that when a fish is caught it sometimes signals to the others about the danger. Presumably what is responsible is a chemical that spreads in the sea and alerts the rest of the school.

This kind of communication is found in insects also. Do you recall our mentioning butterflies that can detect the presence of a female at several miles? There is surely "something going on here"—but what?

It is known that ant societies, which are extremely complex, produce very rich and detailed sounds. This communication allows ants to transmit information and instructions concerning everything from the care to be given a sick baby to defense by "policemen" repulsing intruders and profanity used by the "garbage men." There can be no doubt that a permanent language is operating, transmitted in ultrasound or some other indecipherable signals.

Perhaps the best understood of these languages is that of the bee. Zoologists, intrigued by this astonishingly intricate society, have examined them more closely than they have any other insects.

The gesticulatory communication of bees seems to culminate in the most captivating of all movements: the dance. The ballet of the bee indicates the distance, as well as the place, where food is to be found—without being off by more than four to eight inches. A circular dance "says" that nectar is less than 100 yards away; a figure eight means a greater distance, on the order of two to seven miles in times of scarcity. The vibration of their abdomens indicates the distance that must be covered: The slower the rhythm, the greater the distance.

These are only the broad outlines of an extra-human language to which the Austrian entomologist Karl von Frisch (a 1973 Nobel laureate) and his coworkers have dedicated 30 years of research. This work has revealed the astonishing secrets of another world, a world almost impossible to conceive and as enigmatic to us as would be the inhabitants of an unknown galaxy—another logic, another way of thought, another sensory modality.

But all animals, even the most familiar, pose this problem for us. If we can explore time with words and engage in metaphysical acrobatics, if words can be used in poetry or for inventing atomic bombs, should we conclude that animals are more "beastly" than man? Animals live in the present. And their world is a world of sounds, colors, phenomena, and rituals which, in spite of all our knowledge, we have not yet really entered.

This world coexists with ours; we penetrate it at points, while at the same time running parallel to it. Animals and men do not receive the same wavelength. We have not hit on the correct frequency for picking up their signals. Perhaps we shall succeed in doing so. But one thing is certain: Signals are being transmitted.

EIGHT
Play and Learning

The little lion was three weeks old when he became aware of the world around him. His eyes opened, his ears listened: something was stirring beside him. He stuck out a paw and touched another paw that moved. This delighted him. He turned around and stretched out on his back with glee. If he had been a human baby, one would have said that he was laughing. Then his brother came over to him and tugged on his ear: This is really fun! He returned the compliment.

The big lioness purred softly, noticing that her babies were playing. This was the first time. She wanted to play with them; her tail lightly brushed their muzzles. The babies jumped to grab this new toy.

Their life had begun. Up till then, these nidicolous young had continued to lead a sort of fetal life outside their mother. Deaf and blind, they had existed embryonically. In order to survive, all they had had to do was to keep warm and to eat—a wonderful, carefree life in which everything came free. But for the first time, they were being exposed to something else.

A few miles away, a young wildebeest also nursed at his mother's side. He was one of those poor nidifugous youngsters who do not have time to experience a joyous and carefree childhood. From birth on, he had had to follow the herd on shaky legs. His games already consisted of the activities that would constitute his herbivorous life: flight, galloping, jumping—everything that would help him escape predators—and also mock battles, in which he would meet one of his cousins in single combat, their growing horns clashing. These fights were rehearsals for the real combat in which they would engage, once they became mature, for possession of females during the breeding season.

The little lion was five weeks old when, one morning as he woke up, he was seized with curiosity: a totally new and intense feeling which impelled him to act at once. His mother was sound asleep, her nurslings safely nestled between her paws. The little lion got up and, still a little unsteady on his feet, took several steps outside his stomach-nest. A bird trilled, and in the distance a young hyena yapped exuberantly. The sun caused the shadows of leaves to dance on the ground; the cub tried to catch them, but in vain.

Giving up this game, the young lion headed toward a clump of tall grass, plunged ahead, and suddenly noticed that he was all alone. His whole world had disappeared. Where was his mother? He called for help. A worried cry answered him—it was his mother, who was looking for him. Reassured to know that she was so near, he ventured forth in the opposite direction. How new and interesting everything was!

Suddenly, the cub felt himself being lifted up. His four paws clawed the air and he growled with anger. Then a feeling of pleasure suffused him at the spot on his neck where his mother had grabbed him, and he was appeased. Soon afterward he was deposited unceremoniously among his less rambunctious brothers, and received a well-deserved growl and slap from his mother. He was taken aback by this: He had certainly learned a lot this morning!

He experienced something else that day for the first time: pain. His swollen gums hurt him where his milk teeth were coming through. To soothe them, he nibbled at his mother's paw. She responded by playing with him, and he soon forgot his discomfort.

The lion cub was between two and three months old when he escaped his mother's surveillance and fled toward the unknown that held so much fascination for him. He romped at random—a need that he had to satisfy. He had tried to catch a butterfly and had played with a rodent without fear of harm, when a muffled sound like that of a storm transfixed him.

A herd of gigantic, fearsome elephants passed in front of him, the Old Man in front, followed by the males surrounding the females. A baby elephant had curled his trunk around the tail of one of the females, so as to be sure not to lose his mother. Another young elephant had found this amusing and had imitated him. In fact, there were a half dozen walking in line, each with his trunk grasping the tail of the one in front. It was as much fun for them as playing ring around the rosie is for little humans holding hands. Zebras play the same game, holding their neighbor's tail in their mouth.

All this certainly did not make up part of the young lion's fun. Seeing these huge animals, he withdrew prudently and returned to his mother, who had not—luckily for him—noticed anything. One of his brothers was awake, so they engaged in some play-fighting under the benevolent gaze of their parents. This play was no longer the gratuitous behavior of the initial episodes, but rather was what scientists in their wisdom call "rehearsal play." This gibberish means that, by playing, the young animal enacts his future life: foraging, fighting, sex, etc. Without knowing why he did so, the little lion mounted one of his brothers: Bending over, he was astonished to see his penis protruding from his fur. Very embarrassed, for felines are terribly modest, he recovered by going off a little way and giving himself a bath.

Then he got another brainstorm and threw himself onto the patriarchal beard of his father, who repulsed him with a growl: The lion does not like to be bothered by his progeny. Paying no heed—which was wrong: the lion can quickly turn mean—the youngster bounced playfully over to his mother. But she in turn also pushed him away; she was going after food and had no time to waste. Adult animals are like man: They play only if they have nothing else to do. When the lioness had returned, her paws and mouth covered with blood, she first had to perform her toilet. But then she let the youngsters borrow her tail, so that they would leave her alone. Initially it had amused her to play with the cubs, but now she was a little tired of it.

Parents certainly don't understand! His curiosity aroused by his previous adventure, the cub decided to go farther. Leaving his less daring brothers at home and taking advantage of his parents' preoccupation, he set off into the brush. His mother and aunts had left to hunt and his father was sleeping in the sun—and anyway, this big egotist was indifferent to what his children did.

Something shone in the sunlight: white, round, and heavy, it attracted his attention. He probed it with the tip of his paw without getting it to budge. He applied more force and the object rolled on its side. Another toy for him! Completely absorbed, he pushed the ball and then hid behind a tree, dashed out, and leaped. All of a sudden, a tremendous blow on the head half stunned him. Before him stood an immense, furious bird whose beak pummeled him with the speed and power of a steam hammer. The ostrich was not going to let anyone touch her egg!

Whimpering and crying, the cub fled as fast as his legs would carry him. Life was not as much fun as he had imagined! Nevertheless, his curiosity had led him to a new aspect of play: that of experimentation.

Scientists, with their need to prove things that are obviously true, have discovered that if one gives a stick to young monkeys, a series of experiments follows. After having played with the stick—a random behavior—they examine it: curiosity. Then they try to use it in different ways—experimentation—and finally they discover that it can be used to free a banana—learning.

The young lion and his brothers conducted experiments too. They were six months old when one of their aunts asked them to join her one evening. She led them, jostling each other and clowning around, to an open space that provided an extended view of the savannah. There the lioness began staring at one spot so fixedly that the cubs looked in the same direction too. In the distance they saw their mother and two of their aunts crouching stock still, ready to ambush a young zebra who had escaped from the herd.

Suddenly, with a mighty leap, one of the females attacked the animal, who was soon subdued and dead. Wild animals only kill to eat and never make their victims suffer any more than necessary.

This "lesson"—for that is what it was—was taken up again several times. Eventually the mother lioness decided to take her young ones with her. Night fell. For the first time, they went on a hunt.

Before anyone else could, the cub saw a gazelle. Like a kitten playing with a ball of yarn, he jumped up in the air. He was about to commit himself to the task joyfully when the heavy paw of his mother stopped him with a cuff on the muzzle. Silently but effectively, the huntress pushed him into the rear. In order to be successful, it was necessary to do as she did: to crouch behind a bush and, with patience and vigilance, await the right moment.

These lessons lasted for some months. Learning was taking up more and more time than playing. One day the young lion struck down his first prey. He had just passed a test. This made him an adult lion, responsible henceforth for his actions and his life.

Even though this story was told by a layman, it is not at all anthropomorphic. Most young mammals follow this path, especially if they are nidicolous. They proceed from free play to exploration, then to curiosity, and experience, and finally learning per se, usually instruction given by the mother for which the puerile amusements of the first weeks have prepared the youngster, little by little.

Actually, in the beginning there are only totally haphazard acts, by which the youngster expresses his excessive vitality in spirited exertion. This is the age when even fights are devoid of aggressiveness, in this way resembling the play of the child who points his finger at someone and says, "Bang, bang! You're dead!" without having any idea of what death is.

A little animal's play is a sign of idleness. He plays because he has nothing else to do as long as he is entirely dependent upon his mother. By consigning themselves to man for their survival, both dog and cat amuse themselves in this fashion for a much longer time than do wild felines or canines. In this respect they are just like the "idle rich" who, even though they have reached adulthood, continue to play—at tennis, bridge, love—because they do not have to earn their living.

This sort of play easily becomes make-believe. The feather that the cat flips about with his agile paw becomes a bird, the wad of paper a mouse ... Youka the dog hid her ball and then pretended to search for it: a psychodrama!

But play is always going on; one might even say it occurs more and more in the evolution of a given species. For example, a dog plays at running because this activity (either in flight or in tracking) comprises a large part of his canine life. A little monkey, like a human child, plays with his hands. A cat poised to spring at a ball of paper is anticipating the deception she will employ to surprise her prey.

Moreover, play hones the primordial senses of each animal. The puppy searches for the ball with his nose, and the kitten sharpens her eye by tracking the movements of the feather. To this extent, play refines the inchoate instincts.

Oftentimes, even though curiosity has ceased to play a part in the animal's play, it remains the predominant motivational force. Thus the rat, even though he constitutes one of the rare species that do not indulge in play (at least free play), turns out to be a born snoop. If he is placed in a cage with two compartments, one of which he has already explored and which contains food and the other of which is unfamiliar, he will not hesitate to head for the second, despite his natural gluttony.

The American biologist Nissen (who must be considered a sadist by his rats) made a further demonstration of this. He installed an electric grid in the experimental cage; despite this, his rodents did not hesitate to cross it in order to have a look around!

The cat remains playful to the end of his life. His fascination with the unknown is unlimited. A paper bag, a partly opened cupboard, a drawer attract him irresistibly.

A born mathematician, Yahn was a large blue Persian who was extremely pensive. Play seemed to hold no further appeal for him once he had discovered the "why" of it. If someone gave him a piece of paper dangling at the end of a string, he hit it two or three times with his paw—as precisely as would a scientist conducting a delicate experiment. Then, sententiously, his head buried in his thick beard, his eyes half-closed, he could be seen very clearly to begin calculating. At last, his paw climbing up the length of string to the point where it was held, he struck the string at the top while watching the paper which moved below. Having figured this out, his experiment completed, he was no longer interested. What human researcher could criticize him?

As for primates, nothing interests them more than the behavior of their conspecifics. Moreover, it is thanks to their gift for observation that the young learn from the adults what they can and cannot eat. But what all monkeys like most—even better than their favorite fruits—is to watch a color film depicting ... monkey life! In this they reveal themselves to be terribly human and even literary. One such film was even shown to them upside down (scientists have these perversities) but they didn't seem to mind very much; they merely turned their heads upside down. But then they objected: This position was not very comfortable, even for a monkey!

A chimpanzee lived among people. He thought he had the talents of a handyman. Nothing interested him more than the television set. Combining an observant eye with a keen curiosity, he peered behind the set to understand how the image arrived at the little screen. Then, methodically, he took it apart. Later, discouraged, he abandoned his junkpile.

The sense of exploration is part of a young animal's play. This, along with curiosity, forms the basis of all his activities. No stimulus whatsoever is needed to release it: hunger, thirst, sex drive. Animals love to walk about and observe—perhaps, very simply, so as to combat boredom.

Another animal with a lively curiosity, the horse, will become bored if he is left alone in a stall. He becomes so distraught, in fact, that he may develop aerophagia, like a person, or he may resort to tics in order to distract himself. It is therefore essential to provide him with a little companion to cheer him up by piquing his curiosity. A goat, a ewe, or often a little dog accomplishes the purpose, and a redhibitory defect, "aerophagic tic with or without wearing of the teeth"—strictly forbidden by the law governing the sale of horses—will have been avoided.

Exploration is the logical sequel of curiosity—or, more exactly, the manifestation of it. The cat and dog walk about in order to have something to do but also because they are curious about what is going on outside. Inevitably this curiosity is a way of learning about life. Curiosity will teach the innocent babe that life is not always sweet.

Coming across an egg-toy, the little lion does not yet know

about the beak blow that will follow. Later he begins to be suspicious. It is at this point that play has again changed form, becoming a dramatic representation of what will take place later in life.

The little girl who irons her doll's clothes is learning, without being aware of it, her future role as a housewife. The cat who licks a woolen toy acts in the same fashion. All that is left to the mother is to lend some finishing touches in order for education to be complete.

Youka had four sons, one of whom resembled her so closely, hair for hair, that he was named Sonny. Whereas her other children had left without her showing the least distress, she absolutely insisted upon keeping Sonny, imposing him on her master, and continuing to play with him.

Their play together was strange, for she nipped at him constantly, driving him away or pretending combat to the point where the exasperated youngster showed his teeth and growled at her. Only then did she show her contentment by licking him. She did so tenderly—he had worked hard. Actually, she was teaching this little softy how to bite and defend himself!

Usually it is the mother who teaches the offspring, but it may be the mated pair, as in the red fox—or the entire social group. The young stag, once he reaches puberty, is raised by the males of the herd.

The interest in play of the young—human or animal—is undeniable; It is a fact. This form of education is necessary. But what is the basic origin, the "what," of play? We have no idea. What we do know for certain is its enormous importance in general behavior and also in the tame animal's relations with children.

We also know that play corresponds to the beginning of concrete intelligence. So-called "inferior" animals—the fish, the snake, the bird—do not play (with some rare exceptions). As for the toad, in passing from the tadpole to the amphibian stage he literally has no time to fool around! His metamorphosis satisfies him. Perhaps metamorphosis is a game!

The absence of play does not imply the absence of learning. Birds learn to fly, and even to sing: there are nightingale schools. Before leaving on a migration, birds force the fledglings to make test flights to see if the latter can stay in the air long enough to travel the necessary hundreds of miles. Those who prove incapable of keeping up are killed without hesitation, among swans and other birds.

Should we not also mention the kind of learning that takes place by "insight" when, all of a sudden, the animal understands something that he will never forget? Thus, some urban dogs have discovered the meaning of a red light and use this knowledge in crossing streets, just as people do.

This brings to mind the story of the Japanese monkeys.

On an island surrounded by a current of warm water there lived about a hundred macaques. This troop was blissfully happy because the government had transformed the minuscule island into a preserve. But never in living monkey memory had anyone seen a single macaque bathe in the sea! And if a baby approached too close to the waves, he earned a spanking.

This went on for about 20 years. Then one day a very young monkey dropped a piece of fruit into the water. Caught up in his play and doubtless taking advantage of the fact that he was alone, he threw himself into the sea in order to recover the object of his desire, which was balanced temptingly on the crest of a wave. No one had taught him how to swim but he discovered, as a dog does, that this behavior is innate. The water was made tepid, almost warm, by the current. It was delicious to play there.

Deaf to the shouting of mama and all the rest of the troop assembled on the bank, the little macaque swam around and had a great time: water was as much fun as the land and the trees!

And that's just what he explained (in monkey language) to his elders when he climbed onto the bank again. Several adventurers subsequently followed his example. Today, the entire troop goes swimming. Better still, the macaques have learned to use the water: They wash their food in it before eating!

Another example: near Lake Manyaca in a Kenyan preserve live some lions—kings of the savannah—that perch in trees! True, the trees are inclined, but this is the only spot in the world, except for another small region of South Africa, where lions perch in trees ... like common leopards. Originally, casual play and then learning doubtless created this "custom."

This is where play can lead.

PART TWO

Animals and People

If a cat dies a natural death in a home, the people living there shave only their eyebrows; if it is a dog, they shave their heads and their entire bodies. . . .
—*Herodotus*

NINE
If Dogs Had Hands: Training the Animal

Uépo, a wolf-dog, and Jolie, a black cat, slept next to each other. Having known each other's language for a long time, they understood each other perfectly.

Their masters were in the living room, in front of the television. Just now the cat was watching a film with them, a film about birds. The birds so excited her that she found herself on the floor, having tried to jump from the table, where she had been perched, onto the screen where a gull was flying. Thus frustrated, she went to rejoin the dog in the next room.

Later on, when the television had been turned off, the two animals reacted analogously. Jolie, meowing, hurried toward the door of the bedroom where she was in the habit of spending the night. And Uépo barked in front of the garden door, about to embark on his customary little tour. Experience had taught both of them that, when the television becomes quiet, bedtime has arrived. This was a conditioned reflex in each of them. And even if animal psychology is completely different from our own, it is easy in this case to interpret their behavior. We can understand it.

Ordinarily, once Uépo had urinated and had completed his constitutional, he went back inside at his master's command. But for several days now, as the man waited for Uépo he played impatiently with the light switch that lit the garden: light, dark, light, dark, coinciding with the call "Uépo." This evening, since he had just reread several books about reflexes, the man decided to give in to his curiosity and conduct a little experiment. Several times in a row he turned the light on and off but without calling the dog. And Uépo ran to him promptly, just as though responding to the sound of his name! Thus it seemed that Uépo had automatically replaced the customary vocal command with the signal "light, dark" which had accompanied it in the recent past. This experiment was a demonstration of the famous "conditioned reflex," discovered by Pavlov in 1901, which made that Russian physician world-renowned.

Pavlov made this discovery only by chance because at the time he was still a gastroenterologist conducting experiments on gastric juice. It was known that a piece of meat makes a dog salivate; this is an "unconditioned reflex." Pavlov paired the sight of meat with the sound of a metronome. One day the dog did not have the meat in front of him, but because he heard the metronome and the sound evoked his memory of the meat, he salivated. This marked Pavlov's scientific discovery of the "conditioned reflex." The word "scientific" is important, for animal trainers had utilized this technique for centuries without "knowing" it. Nowadays, the can of dog food has replaced the metronome for the modern dog.

The "Pavlovian reflex" proved to be the first big step toward man's comprehension of animals. Up to that time, despite the studies of Réaumur, Lamarck, and Cuvier, mankind as a whole was not very interested in animal behavior. Ever faithful to Descartes, they regarded animals as "beasts" and reserved their condescending affection for species familiar to them: dogs, cats, horses. As for zoos, these were mainly for visitors: inquisitive people but not really serious, such as the beribboned nannies who took their well-disciplined children there.

Pavlov's experiments were to attract the attention of scientists and subsequently all mankind to the mysterious world of animals. But not until 1932 was it realized that one could go further still, that a reflex could be not only conditioned but also deliberate. That is, just like man, an animal is capable of proposing a hypothesis.

This was discovered at the end of an experiment that had begun in the usual, classical manner. Rats had been put into a cage with two runways, one leading to a dark cul-de-sac and the other well-lit and leading to food and some open space. After having explored the two runways, the rats were quickly conditioned: They chose the one that was well-lit.

The biopsychologists then figured out how to complicate the problem: sometimes it was the dark runway, and sometimes the illuminated one, that led toward the rewards. Which would they choose? The scientists awaited the rodents' reaction eagerly. What would they do—be discouraged, go crazy, run from one choice to the other?

None of these things occurred: they deliberated! Not a little puzzled, the scientists were forced to admit, after some thousands of experiments, that the rat made a hypothesis on the basis of the rhythm of the changes of the runways! The rat had to be resorting to statistical calculations in making his hypothesis concerning which runway to take.

This behavior was all the more remarkable because it corresponded perfectly to intelligence as man defines it: "The capacity to solve new problems by means of thought." To go a bit further, are there so many people who are capable of acting as well as rats in this respect?

Abandoning the white rat preferred for laboratory studies, scientists then turned to animals nearer ourselves, especially the dog. They gave him the same problems as the laboratory rodents, but these hardly caused him any more difficulty. The dog can reason. Very quickly he found the pathway that led to the food or other reward.

The cat has a lot of trouble understanding spatial problems;

He is not very gifted in this way. On the other hand, he can find some answers that neither rodents nor canines can think of.

Enclose a cat in a cage and give him some toys that will interest him. Place some food outside the door. When he gets hungry he will examine all the possibilities for procuring the food, but he will quickly see that nothing will work. Then he will stand stock still, eyes half-closed, and reflect—just as do those barbarians who think up these sadistic games in order to bring themselves up to an animal's level of intelligence!

Finally, having understood that the toys must be crucial to the solution, he strikes the ball tentatively. If nothing happens, he gives this up in short order and tries something else. With his eyes and paw he examines the strings that hang here and there, until he finds the one which, if pulled, will allow him to get his food. This may take a whole day, even longer, but he will always find it.

At first blush this task may seem to be an exercise for training a performing animal; trainers would recognize it as such. But to biopsychologists it constitutes a very serious experiment. Among other things, it demonstrates that the cat thinks before he acts. Dogs and rats, who possess less imagination, never make any connection between the string and the food; They will starve to death first.

On the other hand, the rat effortlessly learns geometric symbols that allow him to obtain his lunch: crosses, circles, triangles, squares. His behavior is identical to that of chimpanzees, except that chimpanzees can carry things further: They can even identify geometric figures that are upside down.

Although the rat seems as talented as the chimpanzee intellectually, he does not possess the latter's hands. Without hands, even with his genius man would still be living in caves. This marvelous tool is doubtless more important even than the eye or sense of smell since it enables one to "make things work." A monkey can take a cigarette out of a pack as well as a man can; a monkey can seize a box and examine it, or open a drawer. His hands enable him to solve problems that other animals of equal or even superior intelligence cannot solve if they do not have hands.

To demonstrate this, place some bananas outside of a chimpanzee's cage. Give him some boxes and several sticks that can be attached to each other. The ape will look at the inaccessible fruit and the various objects that he has been given. He will pause and think for a moment and then pile the boxes on top of each other and climb onto this scaffolding to try to reach the fruit. In vain: They are still too far from his grasp.

The ape climbs down again. He meditates again, and this time he grabs a stick, climbs back up the boxes, and tries to reach the fruit with it. Again too far away!

Third try: he takes all the sticks, examines them closely, and notices that they can be fitted into each other. He does so, climbs back onto the boxes, and obtains his dinner.

This endeavor requires concentration, intelligent behavior, but also hands. This is exactly what happens when, out in the wild, chimpanzees strip a little stick of its leaves in order to fish for termites, which they are crazy about. The dog would surely be capable of comparable behavior if only....

But intelligence does vary with the species. Dogs, cats, rats, and primates can make a thousand detours to find their food, but horses are incapable of doing so. A horse would die of hunger without thinking of making the little detour that would allow him to find his oats. His intelligence is vested elsewhere: He is a born mathematician, as are birds.

The plaster egg which makes hens lay must have been known from the time of Columbus. But why does it make them lay their eggs? It works for the same reason that, if you take an egg from a canary, she will try to lay a new one.

The hen wants x number of eggs. If she finds none in her nest, being rather flighty and prone to forget what she was about to do, she will stop laying at that point. That one plaster egg is her reminder: She will quickly lay others so as to have the number of eggs that she has decided to hatch! Science was left the task of explaining the effectiveness of this practice, which had been employed for centuries by people who appreciated its utility without trying to understand its cause.

A canary has decided that she shall have four hatchlings. Remove an egg. She "counts" that she only has three and hastens to re-lay a fourth. Remove another: She will do the same thing. This will happen with any bird. Experiments on this subject are conclusive: Animals can count.

Scientists have utilized tests in order to measure animal intelligence, just as they do with people. But sometimes they fool themselves. Some of these tests are very difficult, such as the one where you have to identify the blue rectangle that corresponds to the relationship between a red circle and a bunch of dots arranged in various configurations! Plenty of people are not as proficient as animals on some of these. For example, once these tests were given to 250 people and a crow. Do you know who had the best score? The crow!

Experimental discoveries have led to new discoveries, and today we are sure that animals act not only by instinct but also by intelligence, which entails reasoning and therefore independent judgment. Granted, this form of intelligence is not comparable to the creative intelligence of man, but even though it is rudimentary and limited it cannot be denied. It exists, just as there exists a language made up of signals, imitation, gestures, and sounds.

Taken as a whole, these experiments prompted the German ethologist Wolfgang Köhler to affirm that "man and animals are both capable of insight." What differentiates the two types of intelligence is, obviously, the creative drive which is uniquely human.

Many researchers also believe that although both man and animals can make use of information from past experience, animals have no sense of the future. But this is not so clear. Why does the crafty fox play dead (this ruse being, incidentally, proof that he can interpret a situation) if not because he knows that this will allow him to catch chickens—in the future?

And what about the dog who is banished to the kennel by his master who is going away? The first time, the poor thing howls, refuses to eat, chews a gate to bits, fights with the other dogs, and bites his keeper. But then his master returns and the dog understands. Thereafter, when taken to the kennel, he awaits his master's return with melancholy sadness. Past experience has taught the dog that his master will always return. And this return takes place in the future.

The most extraordinary experiments on animal intelligence have been conducted with white rats. These animals have been shown to combine reasoning, comprehension, and cooperation.

An electric grid is installed in front of their food. The rats must first understand that they have to sit on a switch to shut off the current, if they want to eat without getting a shock. They learn this rapidly. Next, the switch is moved away from their feeder, so that the rats can no longer sit down and eat at the same time. What should they do?

This problem was neatly solved by the rodents. One of them sat on the switch while the others ate lunch. When one of these animals had finished eating, he exchanged places with his buddy, so that the latter could get to eat!

It is quite an extension of reasoning, intelligence, and even altruism to have an exact sense of future time and think in this manner, to say to oneself, "If today I do not take his place, tomorrow will he allow me to eat?" In common language, the rats take turns.

If we ourselves occupied the place of the Machiavellian scientists who invented this diabolical procedure, we might ask ourselves what these rats, who are clever and cooperative with each other, think of people!

TEN
A Pigeon to Change the World?

"The face that launched a thousand ships" we say in reference to Helen of Troy. Well, after the Second World War there was a pigeon who might have changed the face of the world.

Actually, this was only a tame pigeon. But his training was very special. The olive branch of this dove of peace was replaced by . . . a missile! The bird was trained in such a way that he could guide the missile toward its deadly end with much more precision than any electronic device.

This took place around 1944. At that time at Indiana University there was a certain Professor Skinner, chairman of the psychology department. His colleagues considered him more of a buffoon than a professor worthy of the name. Had he not taught a bird to play the piano? And rats how to play billiards? Was this sort of activity appropriate for a serious scientist?

But it was his scornful colleagues who were ignorant. The work of the strange professor was taken very seriously elsewhere—"elsewhere" being the Department of Defense in Washington, where it merited Top Secret classification.

The pigeon is wrongly regarded as merely a menace to monuments or a delicacy with petits pois. Actually, he is a very apt pupil for trainers. If he is taught to do so, he easily learns to locate a particular image even if it is buried in the middle of other images, and to strike it right at its center. The pigeon receives a grain of wheat as its reward each time that he responds correctly.

This kind of learning can be accomplished very quickly, and the bird, if he has been conditioned properly, always strikes just the right target. Therefore nothing was easier than to place a pigeon in a missile where, thanks to radar, it could see everything that it flew over on the earth below. Having previously learned what target to destroy, the bird-pilot would strike it when it appeared on the screen, thus releasing a correcting device that would aim the missile directly toward the objective. Certain models of the "Skinner missile" even carried seven pigeon-pilots so as to correct possible errors! Only the end of the war stopped this ingenious project.

The Russians were more advanced in the training of dogs: They taught them how to blow up tanks! Nothing could be simpler. The animals were conditioned to look for a piece of meat between the wheels of the tank. At first, the tank was stationary; later, the tank was made to go progressively faster. At the right moment, a charge of dynamite was placed on the dog's back. Like a good soldier, the dog rushed toward the enemy armor and blew up a tank, and himself along with it. Hardly more than three dogs out of ten refused to obey: the "blockheads."

Nowadays we speak of the "other-directed man." This appelation also applies to animals, because they are conditioned by the same forces: mother, family, society, and of course, man.

This conditioning of animals by man has occurred formally for centuries: We call it training. However, we cannot explain scientifically what motivates this behavior in animals, except to say that their aptitudes, tendencies, and even their shortcomings can be utilized—just as in teaching people.

Hawking the detergent "that cleans by itself," the huckster takes advantage of a virtue and a fault: cleanliness—the linen will be spotlessly clean; laziness—it takes no effort. One acts in exactly the same manner in teaching birds how to count. One takes advantage of their gluttony and of their astonishing ability to count.

This is very effective. If one trains a bird to eat five rows of seeds and gives him seven, he will leave the last two alone despite his great appetite. Conversely, a jackdaw who was entitled to seven seeds made a mistake one day: He took only six. Several moments later he realized his error and returned to look for the seventh!

But of all animal behaviors, the strangest of all, without any doubt, is that of Elberfeld's mathematical horses. These horses lived in the last century in Germany. They had been raised not by circus professionals but by amateurs, and they could add, subtract, and multiply. Incredible? Not at all—a bagatelle! For they quickly gave up these exercises, which doubtless they considered too "juvenile," and took up the practice of calculating square roots!

At once, news of this prodigious feat spread around the entire globe. Some backward people attributed the phenomenon to the devil, but scientists—physicians, psychologists, zoologists—came running to see the "miracle" close up. Indeed they were all skeptical.

One thing was certain, however: the owner of the horses, Herr Krall, was above suspicion. He himself believed in the horses. Moreover, he was the only one who did.

Then Edouard Claparede came on the scene. He was a famous Swiss psychologist who had been sent by his country to conduct an official investigation. For Claparede's benefit one of the horses extracted a cubic root. So here was our scientist, very disturbed, vacillating among various explanations ranging from mental telepathy to the presumed extraordinary intellectual powers of these horses. The man finally concluded that, without knowing it, the audience was tipping the animals off and the latter were picking up these cues—an indication of very high intelligence indeed!

A second investigation took place. A German, Professor Hartkopf, sequestered himself with a horse in a horsebox and gave him a problem to which he himself did not know the answer: It concerned the fourth root of a six-figure number! Ten seconds later, the horse, tapping his two hooves on the ground, gave the exact correct answer!

No one ever found out the truth. No one ever understood. However, this last experiment left open the possibility that the horse actually had done the arithmetic. It is difficult to conceive of him as a mathematical genius, *equus* not being very intelligent. But it is just as impossible to question the reports of the scientists of the time.

So? Mental telepathy? Hypnosis? No one knows. The phenomenon has not been duplicated since.

A young woman of our acquaintance happened to have a counting horse, doubtless trained by its previous owner. One day she noticed that the horse could add. This was quite a modest accomplishment compared with Elberfeld's horses, but all the same. . . .

Without carrying this any further, let us say that a person can easily condition an animal simply by making use of its normal behavior and its acquired reflexes, including its conditioned reflexes.

However, the most extraordinary examples of training are not achieved by scientists (who have a predilection for the unpretentious white rat) but rather by circus trainers, whose preferences run to large wild beasts. Incidentally, one wonders why the study of large wild animals has been neglected by biopsychologists.

Professor Hediger, director of the zoological garden of Zurich, has taken upon himself the task of translating into scientific terms all the practical training techniques that have been handed down from father to son. This is a backwards sort of task: given the solution, find the question!

Hediger has found that a captive animal's cage constitutes its little territory, with all that this implies. Thus, the trainer's stool represents the tiger's core area. If the territorial perimeter is drastically reduced, it is all the more staunchly defended. If anyone passes beyond a certain point, he will have a fight on his hands. But several inches outside that area, threats will suffice. The trainer is fully aware that he assumes no great risk if he stays beyond that line, keeping the animal on the defensive. The wild beast makes menacing gestures: simple bluffing whose only effect is on the spectators.

It is well known that animals living in groups or families usually have a leader and obey him. If a man is recognized by them as their leader, they respect him and accept his commands.

To achieve this, the animals must still be quite young when they are first trained. Further, the trainer must understand their "language" and their "laws," which they insist be upheld. Thus the trainer will always be the first to enter the cage, thereby assuming possession of the territory for his purposes: Everything that he does should suggest authority and inspire respect. Then, self-assured, he can converse with his "animal family" by means of signs, gestures, or sounds: a code understood by all the animals. Once he has achieved this degree of familiarity, the trainer can demand a great deal of "his own kind," who will do their utmost to please him. For this reason the training of animals living in social groups—lions, primates—is much easier than that of solitary animals such as tigers and panthers.

By the same token, horses are easily trained, provided that one knows how to utilize their innate tendencies. The famous school of Saumur, like that in Vienna, is as dazzling as it is only because the horsemen are perfectly familiar with the natural behavior of horses and know how to condition them.

They know, for example, that a frightened horse gets up on its hind legs. Taking off from this fact, in order to trace a figure, they evoke a conditioned reflex in which the animal's fear is replaced by the rider's command.

Trainers have known all this for centuries, at least on the practical level. They might have been able to teach this to scientists who, alas, have had trouble getting over the simplest animal behaviors and have therefore drawn some astonishing conclusions.

This was the case with Pavlov. Moreover he declared it with all the candor of a true scientist:

> In the course of our work we found ourselves at an impasse at one point; we had not succeeded in understanding what had happened. ... For our experiments, we had a very intelligent dog who quickly entered into amicable relations with us. At the beginning the dog remained calm, eating with pleasure. But the longer the session lasted the more excited he became. He wanted to get away; he scratched the floor, gnawed on the table, salivated ... etc.

This lasted several weeks and left our scientists very frustrated, for their work was interrupted. They were also very perplexed: What could be the trouble with the dog? "We entertained numerous hypotheses on the possible causes of this condition but none succeeded in explaining it...." Nevertheless, Pavlov stated naïvely, "we already possess a great deal of information about the dogs ... ." What could the explanation be?

Finally they found it: The dog had a strong aversion to being tied up! Thus Pavlov "discovered" scientifically what he named, so as not to lose face, the "freedom reflex!"

Pavlovian conditioning is a passive reflex.

It was just in 1962 that the psychologist B. F. Skinner forged an important link with biopsychology by discovering that an animal can be conditioned voluntarily. Based on this, the notion of "reinforcement" was introduced (empirically, the utility of rewards had always been recognized). At this point, obedience became the means of attaining a desired goal: "If you are good, you will get the candy." The reinforcement could even be symbolic: "Good boy." And the dog did what he was told, in order to be praised.

Youka the dog, a film star, knew that she was entitled to a cake when she was finished shooting, but this material reward quickly ceased to interest her. On the other hand, she loved being congratulated. A "great performance!" uttered in a certain tone filled her with joy. On the other hand, if the director announced, "Let's shoot it again," she would repeat the scene entirely on her own and without anyone saying anything else to her. For that dog, a born comedienne, reward and punishment were entirely symbolic. This is often the case with hunting dogs.

Voluntary conditioning can seem to disappear if it is interrupted for a long period. But this is only an illusion.

Several years ago, a circus in financial trouble had to sell some equipment and animals, among them three trained elephants, which were purchased by the zoological garden of Lyon. There, housed in a tight space, the pachyderms trained themselves to extend their trunks in order to obtain a tidbit.

Four years later, their old trainer passed through the city and went to see them. He found them completely stupid. He asked to be allowed into the enclosure. The keeper began screaming. The director of the zoo refused: The animals were considered dangerous!

But by insisting, the trainer finally gained permission. He entered the enclosure. The elephants did not even seem to recognize him. He headed toward the oldest of the three and spoke into the animal's ear as he had been accustomed to doing. At that instant the enormous gray mass, which up to then had continued to extend his trunk mechanically toward the spectators, turned toward him, trumpeted heartily, and waved his ears. Alerted, the second pachyderm arrived in turn. Only the third, whose training had only just begun when he was sold, was apparently disinterested in what was happening.

Suddenly the elephant leader waved his trunk toward the man and wrapped it around him, but without showing any hostility. This was a sign of friendliness. Actually, it was much more than that. Subsequently the trainer would explain, "He asked me to have him perform his old number." And none of the people who were present at the scene expressed any doubt about his interpretation, as surprising as this may seem.

In fact, as soon as the man had given his commands, the elephant, who was immediately imitated by the second, resumed without hesitation the work which had been interrupted for four years. And as the third one remained disinterested in his fellows, they went after him and made him work with them!

This sort of conditioning at man's direction can, at any given moment, be supplanted by the animal's will. This is especially true of a domestic animal such as the dog. If the Greek philosopher Epictetus "wanted what life demanded," the dog wants what his master desires. Trained to attack, he willingly attacks; and the same is true of a hunting dog.

But, as we have seen, man does not act alone in conditioning an animal. Many other factors enter in; for example, animals are prone to generalize, beginning with man himself. Thus, like the Englishman who lands in France and, seeing a redhead, deduces that all French people have red hair, a cat who is chased by a dog concludes that all dogs will chase him. In this feline, who is thought to be the animal closest to man mentally, this attitude can quickly lead to a neurosis!

Of all behavior patterns, there is one that doubtless was originally conditioned but probably is biologically based today: the behavior of the domestic animal. Our pet dog or cat is really only a member of our family because he has been a part of our family—one of its members—for millennia.

ELEVEN
People and Their Animals

Noon, the month of August, a beach in Greece ... and a little blond dog who whimpers, lifting his paws, one after the other, which have been burned white by the sand.

"Oh, how I love animals! Oh, I want him! Oh, look, he is coming toward me ... come, you poor thing, come!" The young woman leans over the cocker that has placed his forepaws on her clean slacks, eyes and voice pleading for protection against this burning pain.

"I have found him; he's 'my dog'—I feel it, I know it. He was made just for me! I told you that I wanted a dog; well, here he is—he was waiting for me!"

The man looks at the woman and the animal, a little annoyed by this excessive demonstration of tenderness by a woman who is no longer a little girl, and who a moment ago was not so passionate. "You know, Monique," he says cautiously, "he's pure-bred; he must belong to someone, he must be lost. We ought to ask the bartender; maybe he knows the owner."

Monique cradles the cocker in her arms and shows her friend the beatific, suffering face of a martyred saint. "I don't care! He's my dog!"

What can one do in the face of such ardor? Well, one can nod to her respectfully, as did the young man who had just arrived on the scene, and explain in halting French, "He escaped this morning. I've looked everywhere." And he held his arms toward the cocker whom Monique held to her bosom as though it were her own heart that someone wanted to snatch away.

The astonished young man looked at the annoyed man. The woman said, "Men understand each other. Explain to him that he's my dog."

A conference takes place for a quarter of an hour before the peaceful sea. Several large banknotes pass from hand to hand. Monique's companion turns to her with a smile: "He's nice, this fellow; he is offering you his dog since you are so attached to it." She answered him with such a smile of gratitude that her friend refrained from saying that for what he had just paid he could have bought her three cockers in Paris.

What he still did not know was that, between the homesick dog who did not understand French (and how!) and the lovesick woman, the return trip would be a purgatory that he would remember for a long time. They returned to Paris. Several days later, Monique telephoned. "I adore the dog! He's wonderful (sigh). But what a bother! Worse than a lover! Think of it: I have to walk him three times a day or else he'll make peepee in the house. And since I don't dare leave him at home alone, I'm afraid that he'll wreck my chairs! So I took him to the racetrack. Naturally I had to leave him locked in the car, but he cried!" (She stayed at the track from 11 A.M. to 6 P.M.) "Well, all that doesn't matter." (To her at least.) "He's a fantastic dog."

Another telephone call: what can she do to stop the darling animal from barking, jumping, yelping, urinating...? And several days later Monique, who loved the darling cocker more and more each day, asked her friend if he happened to know someone who might take care of him because—she was in despair over it—he was impossible to keep: im-pos-si-ble! He didn't know anyone? Well, he was the one who had offered the dog to her; he should shoulder his responsibilities!

So Cocker went back and forth several times: three days at the cleaning woman's house, a week at her friend's, 48 hours here, 24 hours there. For this little homeless dog, life must have become a hell. Monique compassionately delivered him from this sorry existence by having him put to sleep. How she did cry!

This is not just a story: it is a verifiable news item. Every year there are several thousand such occurrences.

For example, there was the cat whose kittens had been kept one autumn "to amuse the kids." November came and the summer cottage was closed up. The cat and kittens were thrown outside to manage for themselves. Winter came early that year. One morning the neighbor heard meowing and went outside. In the snow near his door were three kittens, and the mother was coming up with the fourth in her mouth. Then, having done everything she could to save her brood, she fled. She was found, dead and frozen, several days later in front of the closed door of her "owners."

These incidents are not melodramas told to arouse sympathy, but real crimes. Every wild animal is responsible for his own life, but man, and man alone, is responsible for the life of a domesticated or pet animal.

This is why, before "adopting" a dog, a cat, a bird, or a goldfish, one should understand that he is accepting a responsibility and will have to face up to certain duties, with all that this means in terms of inconvenience and even disagreements. A pet is not just a toy that can be broken and discarded: It is a living being. He has complete confidence in the family that he has just joined—a family comparable to a baby's and on which he is relying in the way a baby relies on its family.

It is traumatic to banish a pet from his family, monstrous to martyr him. To kill him is neither more nor less than murder. Today, in France alone, 10 to 20 thousand animals are abandoned each year at vacation time.

It is no excuse to say, "He's only an animal; he has nothing in common with man." What is supposed to be the big difference between them? The kitten who looks at us with astonished eyes and the puppy whose awkwardness makes us laugh have chromosomes like ours. Their inheritance is subject to Mendel's laws: If an offspring has blue eyes whereas its parents have brown ones, it's because of the grandmother. And if a puppy is blond whereas its mother is black, the grandfather is responsible. Nothing differentiates animals from man—neither organically nor biologically, nor (up to a point) intellectually.

Certainly man possesses the creativity that makes him distinct. But does this justify his imposing tyrannical and "inhuman" treatment on those who are his "companions"—different from and subordinate to him but not so inferior?

The dog, man's companion since prehistoric times, often cannot do without him. Although cats that have become feral again can get along on their own, many a dog will quickly die if he does not find a human companion.

Evidence of the first dogs appears at the same time as that of early man. The dogs did not live with these humans, but only near them. Living outside the settlement, they entered it— furtively—only to pick up scraps on which to feed. (At the beginning of this century, the pariah dogs of Constantinople were still in charge of ridding the city of its garbage.)

When prehistoric man had exhausted the parcel of land on which they lived, they left it for another one that was virgin and fertile. And, most remarkably, the dogs, which were already faithful, accompanied them. It is easy to imagine one of these humans extending a bone to them one day, in the first gesture of friendship. It is easier still to imagine this gesture being made by a child. This must have been relatively early, since the large jars which served as coffins in the Neolithic Age always contained the shepherd and "his" dog, side by side.

As for the cat, he is perhaps more man's friend than man's companion. He does not live for his master but rather beside him. This, however, does not mean that the cat is any less attached to man than is the dog.

The cat's origin is somewhat obscure but is doubtless some wilder feline. So numerous are his avatars that he seems mysterious and disquieting. A black one suggests witchcraft and an incarnation of the devil. On the other hand, the Egyptians perceived the playful cat as the goddess of joy, Bast, daughter of Ra the sun god. "In order to pray to her and honor her, the faithful went to her temple by the hundreds of thousands," recounted Herodotus. Barks plied the Nile to the sound of flutes. People laughed and played, and bantered with the river people. Masquerades, dances, and songs followed each other closely—it was the biggest occasion of the Egyptian year. And when a cat died, he was ceremoniously interred in the shade of the goddess's sancturies.

The Abyssinian perhaps descends from this sacred cat. The sculptural Abyssinian, with his reserved nature, is devoted exclusively to his own master and disdains other mortals.

Man's third companion, the horse, made his appearance in the Tertiary Era about 50 million years ago. He succeeded the giant reptiles, which were on the road to extinction. At that point hardly larger than a terrier, he was to grow—very slowly—with each passing millenium. It was to take 20 million years for him to reach the size of a pony.

The horse was wild and free for a very long time. To find him domesticated, we must wait until ancient Egypt and Persia (3000 B.C.).

Fish, birds, parrots have only become a part of our lives during our era. The noblewomen of the eighteenth century liked having near them a little black page boy and a marmoset. Then the East India Company introduced them to that marvelous discovery, talking birds!

Anthropoid apes may have been domesticated. Prehistoric man may have made them his slaves! At least, this seems to be implied by discoveries made near the border of Tanzania and Kenya, in the Olduvai Gorge.

There, in a volcanic fault, vestiges of encampments have been found. Next to the remains of the human occupants, but not mixed in with them, have been discovered the skeletons of other anthropoids. By means of painstaking studies and research, archeologists have been astonished to learn that the latter beings may have been the workers of the former. The *Hominidae* invented flint tools which the *Anthropoidea* may have used.

From cranial and skeletal measurements, it can be surmised that the anthropoids were on the border between simians and man. This is not so extraordinary if one remembers that gorillas, orangutans, and chimpanzees utilize primitive tools very capably—tools which are comparable with those used by the first men millions of years ago.

The lion and the tiger purportedly rendered a "service" to the Romans in ridding them of the Christian martyrs. Only recently have they been admitted into our living rooms, and it is debatable that this represents any improvement, since the undertaking always ends tragically—for man or for beast.

Once a cute cub becomes a 400-pound lion, he is no longer such a great playmate! A big feline remains a big feline: He lives in Africa and should remain there. If one is determined to get to know him outside his cage, there are plenty of preserves for that and travel is easy nowadays.

In the interest of wild animals themselves, you must not let yourself succumb to the charms of the baby that an unscrupulous salesman is pushing for his own commercial benefit. Furthermore, the more soft-hearted people there are to buy that darling lion cub, the more lion cubs he will order.

Here is a typical story, that of a little leopard that I knew well. He was captured in a tragic episode for which merchants of animal flesh were responsible. In Africa, little animals are crammed into cages, if not into sacks, and often do not survive this treatment. As many as 60 percent have been lost by the time they reach Europe. But this doesn't matter to the seller, given the price for which he can sell them—as long as the survivors can be sold quickly before they die too!

A music hall performer stopped in front of a pet store. A sickly little leopard was exhibited behind the glass. The man looked at this little dear with the innocent eyes and thought that he might be able to raise her and create a world-famous act with her. It would be like a child's dream come true!

He entered the store and asked about the leopard's sex and age. If it were a female, which he thought would be more gentle, and if it were not too young, he would buy it. The salesman maintained that the leopard was just what he wanted: A four-month-old female who was weaned and could live without her mother.

All this was totally false. It was a male and no more than 1½ months old. But if one worried about small details, one would never sell anything! The customer must be satisfied, right away! Afterward, the salesman has nothing to do with whether the beast lives or dies.

Here, then, was the little hero of an adventure that had nothing in common with those that he would have known in his native land. He rode in a Mercedes at 100 mph, making a trip from Nice to Paris. Fifty people petted him, passing him from hand to hand. Nothing is more adorable than a baby feline. He ate at any old hour and anything at all. But his master's affection could not equal his mother's silky belly, next to which he slept so soundly. This little animal is not nidifugous: He is nidicolous and therefore not completely formed or ready to survive. Normally he exists in relation to his mother and cannot survive alone.

This treatment left the poor little thing with osteoporosis, which the man who had taken charge of him, moreover, did not notice.

So here is the drama: Upon leaving the hotel room where he had spent the night with his master, the animal found himself muzzle to muzzle with a boxer. "Hmm, a big cat," thought the dog, who had never heard anyone mention leopards, and rushed after him. Terrified, the poor animal leaped down the staircase. His bones, rendered fragile by the osteoporosis, did not hold: 14 fractures!

When he was taken to the veterinarian, only one diagnosis

could be made: he was paralyzed for the rest of his life and would need constant care.

The actor was in despair; he loved his animal very much. But what could he do? He couldn't take care of him; he was always on tour. So ... he told the veterinarian to kill the animal.

Because he did not want to be an accomplice in this crime, the vet kept the leopard at his home. One morning the animal had a convulsion, stared at his people, and ... left for the Happy Hunting Ground.

Have no illusions: This is the risk you run in buying a baby wild animal. The best possible outcome is to release it in a preserve at 18 months.

The first cousin of the wolf, the dog shares his social characteristics. And the cat, a smaller version of some wild feline, behaves very much as the large cats do. But both species have been accustomed to living among people and have altered their habits accordingly.

Born in people's homes and raised by them, the dog and cat transfer to man not only their affection but also their social and family ties. In the bosom of the human family they have a father, a mother, brothers. ... And dogs have a pack leader— their master—whom they obey not because they have a slave mentality, as some disparaging people say, but because of their sociability. It is no less foolish to scorn them for this than to reproach ourselves for having a president or a foreman!

For an analogous reason, male animals generally prefer women whereas females seek out the company of men. Some of our friends have a female cat who loves her mistress very much, but this does not prevent her from rolling on the floor with pleasure whenever a man comes to see "them," or from acting voluptuously—for a cat—even though she is spayed.

This does not constitute any special behavior in the cat or dog: The majority of the higher animals behave the same way once they are domesticated. Konrad Lorenz has demonstrated this with greylag geese. But it should be added that he was more astonished than anyone else to find that he was the "mother" of Marina the bird!

Lorenz was present when the egg hatched, so he was the first being whom the gosling saw. Thinking that he was doing the right thing, the scientist placed the gosling under a goose. It was at this point that he realized that he was the "mother":

> ... I heard a soft cry, uttered in a questioning tone: "Vivivivivi?" The old goose answered reassuringly but in her natural voice: "Gagagaga." However, instead of being calmed as any reasonable gosling would be, mine quickly came out from under the warm plumage, stared at her doting mother, and then suddenly burst out crying. ... The tiny little thing drew herself up to her full height and continued to cry up a storm. She stopped half way between the goose and me. I made a slight movement. The tears stopped abruptly. The youngster headed toward me, her neck stretched way out, and she greeted me with an ardent "Vivivivivi." This was very moving but I had no intention of transforming myself into a mother goose. So I took the gosling, put her back under the goose's stomach, and began walking away. I hadn't walked more than ten steps when I again heard behind me: "Pfuhp ... pfuhp ... pfuhp. ..." The poor little thing was desperately trying to run after me. But she couldn't keep herself upright yet. She remained squatted on her webbed feet and tottered ahead slowly. Even someone with a heart of stone would have been touched to see the poor little thing running after me crying, stumbling, and falling head over heels. Her mother was not the white goose, but I!

Having adopted the gosling in this way, Lorenz baptized her Marina:

I spent the remainder of the day the way a good mother goose would have. The little thing eventually convinced me that, at least for the time being, it would be impossible for me to leave her alone, even for a minute. If I tried, she was seized with such anguish and cried in such a heartrending manner that I resigned myself to her mercy and fashioned a little basket which I could carry with me constantly....

Night fell and Lorenz ensconced the little bird in a "magnificent, electrically heated cradle." Foo on this cradle! Marina didn't want that, she wanted her "mother." And, after a sleepless night, the kind Lorenz decided:

> ... at 2:45 I made an important decision: I carried the cradle to the head of my bed. At 3:30 came, "I am here, where are you?" (the human translation of 'vivivivi'). I answered in greylag goose language, "Gagaga." " 'Virrr,' " replied Marina, " 'I'm going to sleep now, good night!' "
> But at dawn this was no longer working. Marina began to cry again: What do you do when your dear baby begins to bawl at 4:30 in the morning? You very tenderly hold out your arms and carry her into your bed, all the while beseeching heaven to let the little angel sleep for at least another 15 minutes....

Lorenz concluded,

Marina was truly a wonderful little youngster. It wasn't just capriciousness that she did not want to remain alone for a single instant; it must be remembered that a gosling is condemned to death if it is separated from its mother and siblings. Biologically it is perfectly reasonable for a deserted offspring to forget all about eating and drinking....

Lorenz was thus faced with an established fact: One must not abandon a little goose or else "one runs the risk of seeing it literally die from crying."

Thus, by becoming the "mother" of a goose, the Austrian ethologist discovered the phenomenon of "imprinting," one of the principal observations of animal psychology and the most important of recent years. Imprinting, which operates in all animals, explains why an animal is moulded after the fashion of its master, especially if it is born in his home. If you desire the "adopted" animal to be happy and problem-free, you must be familiar with this phenomenon, and recognize and respect it. If you do, you will provide the animal with an essential family life, and will be provided in turn with an ideal companion.

If not, the animal will be "badly brought up" and, as an adult, will have all the complexes of badly raised children: timidity, aggressiveness, shyness, disobedience, disrespect for authority. He will be a "dumb animal," only it will be the person in charge who is responsible.

This is true of all animals. A parrot who does not feel himself to be part of the family will not talk and will even go so far as to refuse to eat. As for fish that are neglected too long in their aquarium, they will get bored ... often bored to death!

TWELVE
The Dog and His Master

Never before have there been as many animals living with people as today, just as there have never before been as many gardens. These two developments reflect a common need. In our cement universe, in the "dead end" that an apartment is, in a world where the atomic bomb has replaced God as the ultimate fate, and where everything is at hand, wrapped in cellophane, packaged, and refrigerated; in this wonderfully civilized Western world where nice, clean cancer has replaced horrifying plagues, man, rather than being happy, is in anguish. He feels himself very insignificant amid all these great things, for he has not yet reached the stage of thinking of himself as a cell of society rather than an individual.

So, in this universe which suddenly seems artificial to him, surrounded by dishwashers, refrigerators, and automobiles, now that he knows that the supply of oil is limited, he is becoming sentimental and nostalgic. And he cultivates a little blue flower in a corner of his heart.

This can be a clump of rosebushes, a patch of lettuce, a turtle, a dog or cat, or a goldfish. Whatever it is, it will be something stable that has not changed for millennia, the link for which he has an abiding need so he will not feel lost in the middle of the cosmos but will rather feel himself a man among a million years of men.

Many people must share this anguish because in 1975 there were 30 million domesticated animals in France alone:
14,500,000 dogs and cats,
4,850,000 birds,
5,800,000 hamsters,
3,000,000 goldfish,
1,500,000 turtles!
This list does not include the horses, monkeys, mongooses, or various wild animals—lions, tigers, cheetahs—which nowadays reflect less of a real need than a need to feel fashionable.

For most people, to own a pet means to have a dog or a cat. We have understood the dog—in the scientific sense—only since about 1920. To be more accurate, we are now beginning to understand him. But what do we really know about him? Only that he is a canine with all the characteristics of that family, has been man's faithful companion since prehistoric times, possesses extraordinary hearing, poor vision, and a subtle sense of smell, and can be conditioned as readily as ourselves.

All this is rather abstract, and we have a tendency to forget it when we find ourselves face to face with this playful, four-footed rascal that will play all sorts of tricks on us shamelessly and that, from the very first (especially if he is a wolf dog), will try to gain first place, that is, the position of pack leader.

As we have seen, every young animal that joins a human family will make it his own and will transfer all his familial behavioral tendencies to it, to the point where the human smell becomes that of his "pack." This is why one should adopt a dog just when it is weaned, between 5 and 12 weeks, which is a critical period for a puppy.

Up to that point, all he has needed was his mother's milk and warmth. During his first 12 days, still blind and deaf, he has continued to lead the dull existence that he led in his mother's womb. Then, little by little, he has discovered life and its surprises, but always while under his mother's protection. He has had nothing to fear and has known nothing of danger. He has experienced this emotion no more than he has experienced solitude—for he has lived among his brothers and sisters.

Now, at two months, at the very moment when his development is to begin, when his life begins to become external, "someone" takes away his maternal milk and warm belly. And this mother of his leaves him to devote herself to other matters! Forlorn, the puppy feels frightfully abandoned. This is the moment when, seeking his family and feeling hurt, he is searching for someone to whom to give himself, someone who will cuddle him and rescue him from his isolation. If a human who wishes to become his master comes along at this moment, everything will be all right. The dog will pass his youth normally, in the bosom of his new "pack."

This canine melodrama takes place at a kennel or, even worse, in a pet shop. Puppies on display have a pitiful look because they are orphans. Just like human orphans, they too have frustration complexes. And if they wait too long for a master, these problems can become neuroses. For, even more than being imprisoned, what the puppy dreads most, with his well developed sociability, is solitude.

One's heart sinks to see these forlorn and pitiful babies as they extend their muzzles toward man's hand. However, sound judgment demands that one choose preferentially an animal who has just left his mother or, better yet, who is on the verge of doing so. Like Lorenz with his gosling Marina, we will then truly be the "father" or the "mother" of our "baby dog." Let's be clear on this. I'm not referring to being a "da-da to a bow-bow," but to a biological relationship.

Moreover, in order that there be perfect understanding between the future companions, one must call on what Goethe called "selective affinity": one must choose his dog of course, but he must also allow the latter to choose his "human." An animal must never be bought according to the dictates of fashion, but rather on the basis of the affinity that one feels toward him. Every person instinctively looks for his own qualities and shortcomings in "his" pet. Therefore it is necessary—especially in the case of the first "baby puppy"—to inform oneself about the nature of the desired breed and, whatever breed it is, to be completely decided on it. If not, at the end of six months or a year, one will notice that he has not reached an understanding with the animal. Of course, this is not as serious as a bad marriage, but it will still be significant: Divorce is always an unpleasant experience.

Having decided on the breed, you must now let the dog choose in turn. When you enter the door, you must always select the one that comes toward you, even if he is less handsome than the others, because he has chosen his man, his father, his god. The unsteady steps that take him toward you are the first signs of love.

The puppy is now in his new family. He must immediately be helped in getting organized. But his "routine" must suit that of his owners, or else....

The apartment, the house, and especially the yard are his core area. But he also needs a territory with which to be familiar and comfortable. If left to his own devices, he will replace his birthing basket and his mother's belly with the bedroom and bed of "his parents." In so doing he would hierarchize his "pack" and he would become the leader.

It is best to combat the possibility of this catastrophe, starting from the very first night, by teaching the puppy who is going to dominate whom and who is allowing him to occupy this wonderful territory. This will entail a lot of groaning and maybe even some disagreements with zoophobic neighbors, as well as the anguish of the family who has to bear listening to the crying of the little dog who is locked in the kitchen, hallway, or bathroom. But it is necessary to be firm. It constitutes the first show of force, by which the puppy will learn that he is to obey and not command. And there will be others! However, these first days will determine future relations.

While expecting the dog to conform with our mode of life, in return we should respect his social structure, understand his language, familiarize ourselves with his behavior, and recognize and appreciate his hierarchical propensities, which are very similar to those of the wolf in its pack. As we have learned, the wolf pack is led by a male that won his position in fair combat, imposing his will by force but without unnecessary cruelty on a second wolf, that in turn dominates a third, etc. The same thing happens in the human family, in which the dog occupies a subordinate position. He will tend to consider as coming after him any babies, old men, and other animals.

Youka, the famous wolf dog, at last came to recognize—not without difficulties—a human being as her leader. But she herself became the leader of her "pack." She actually had a male under her command (albeit her devoted son, whom she had taught), two female German shepherds, and also—what was even more astonishing—two cats whom she had partly raised and who, acknowledging her superiority, obeyed her.

Whenever the two enemy German shepherds squabbled, the big, black dog would arrive calmly. Without hurrying, without stopping, she snapped to the right and left—biting only thin air—and then walked on, certain that order had been restored.

Furthermore, since she had a sense of heading a pack, Youka insisted that everyone be home in the evening. If one or two cats tarried at dusk, which happened often, she would go to the garden gate and bark characteristically—orders, really—which these free spirits seemed to accept in most cases; they soon returned home, unrepentant like all cats but obedient nevertheless.

The puppy is going to grow. (Whether he will also grow in comportment and obedience is more difficult to say!) He will become a happy baby and will play with whatever comes within range of his paw or teeth. This is his apprenticeship, but it is also biologically based play, since he needs a prey-doll that he can chew on in his fashion: take it by the neck and shake it, as his ancestors did for centuries in order to kill real prey for food.

The dog's owner should also be sure to give him a slipper of his own, or else he will help himself to one. In sum, by knowing the dog perfectly—historically and biologically—his owner can avoid being disobeyed.

But the dog will disobey all the same! A dog who is too well-behaved is sick or abnormal. So the first spanking will come, and it will be followed by plenty of others, up to adulthood (8 to 12 months)!

Punishment should always be fair—not in human terms but by canine standards. Otherwise the dog will not understand and his behavior will remain out of line for the rest of his life. Never, ever should he be hit with the hand, which is reserved for petting only. Moreover, the "naked ape" may hurt himself seriously by beating an animal whose flesh and bones are as hard as his own. However, an occasional couple of good smacks with a rolled newspaper will not do any harm; they constitute a half-affectionate call for order. The dog will use the occasion to make you feel remorseful by rubbing his tingling cheek with his paw. But this must not be taken too seriously: The dog will interpret any show of compassion as a promotion!

The dog will readily accept his punishment if he knows that he deserved it (which will not stop him from being naughty again!). He will even seek it out! On the other hand, he should be pardoned afterward. The gesture of pardoning plays an important role in his education; it should be enacted to the full and never be neglected.

It is easy to tell when a dog has done something foolish—even when you have not seen it. He will betray it by his bearing: head down, body cringing, ears lowered. His master has only to search for the impropriety . . . and punish it. This is what the dog expects. He will not understand if the punishment does not take place, and it will seriously disturb him.

Some dogs will punish themselves when they have done wrong.

Uépo, for example. This shepherd has a passion for hyacinth bulbs. In the autumn he spies on his mistress while she plants the bulbs in the garden, and then he digs them up and eats them! The first time that she caught him, to punish him she locked him up in the tool shed for the rest of the day.

Some days later, she found him lying on the floor of the shed. She had not noticed anything but she understood by his actions that she would not have any hyacinths come springtime! Just like a person, her dog practiced self-punishment.

A memory as effective as this one will not only be used by the animal against himself; it can also act against others. Although the dog is not as spiteful as the horse, he will not forget a wrong that is done to him.

This same Uépo growled and then ran away when "his" veterinarian came to see his owners, even just socially. He had not forgotten several painful treatments. But the "doctor" inspired in him a healthy respect, just like a pediatrician does with children, and Uépo did not dare to bite him.

By the way, this is what happens with most dogs that are taken on a visit to the veterinarian. Furthermore, putting them on an examination table intimidates them.

The puppy will try at least once to bite his master. This is normal; he wants to try to be the boss. But one must never condone a bite, even an insignificant one. To do so would be, according to the law of the pack, to acknowledge to the dog that he ranks above "his" owner in the hierarchy. Then the day would come when he bit his master seriously, not expecting that his underling would object. In this case there would be only one recourse: to get rid of the dog. Whenever you hear that a German shepherd has bitten his master seriously, you can be sure that the man is responsible: He has allowed the dog to dominate him and take his place. Most likely, he was afraid of the dog.

When a puppy cuts his teeth, he bites without a second thought. This is of no consequence. But if no one says anything, he will continue to do so, as a form of play. Therefore one should push his forepaw into his mouth and close his jaws on it. He will whine and be very unhappy, but he'll understand. If he starts up again, one should, without hesitation, seize his neck with one's teeth, as close as possible to his throat! This will teach him who is the leader.

The most redoubtable test of strength between man and wolf dog takes place when the animal becomes sexually mature and defines his territory by lifting his leg (which he does not do until this stage). At this time also, he will try to become "top dog."

At this time the animal is 12 months old. He is a large dog; standing on his hind legs he easily reaches 4½ feet and is powerfully built. His jaws are formidable. This is the stage when his early training—his training as a puppy—will prove to be important. If the dog has been taught never to bite, he will confine himself to play-fighting. But if he has been allowed to bite and hence to get his way, he will fight in earnest and his master may not wind up the winner. In any case, the owner should not take a chance on finding out.

Thus there was a test of strength one day between a man and his dog, a remarkably intelligent Belgian shepherd that had just reached sexual maturity. The dog's tentative biting during his puppyhood had been immediately rebuked, so the animal was subordinate to the man.

This was of no consequence. The trouble began one morning because of some breach of conduct. The man punished the dog severely with a riding whip. Suddenly he saw the open jaws about to close on his wrist. Letting go of the whip, he quickly slapped the dog's dangerous muzzle hard. Then, taking up the whip again, he gave the shepherd a beating such as he had never received before. The dog growled and bristled but did not dare to attack. It was apparent that he had abdicated. His master then locked him in the bathroom, and returned an hour later, ready to pardon him. But what did he find but absolute carnage! The dog had destroyed everything that he could: the curtains, the towels, the bathrobes, even the soap!

Another thrashing. The dog growled, the master responded more sternly—this was truly a case of two strong personalities clashing. This time the animal was tied up on so short a rein that he could not budge. So he chewed up the rug.

This battle lasted till evening. Then the dog submitted: He lay on his back and presented his belly and his throat, in the time-honored manner. His master—which indeed he now was, the dog having accepted him as such—immediately petted and praised him. The dog wagged his tail. Everything was fine, everything was as it should be.

The dog lived to be 18 and never again tried to "buck authority." There is a very simple reason for this: The authority of the pack leader is put in question when he begins to grow old. But man lives much longer than the animal and it is the latter who reaches senescence first.

For the same reason, an elderly person should not get a young dog, at least a large one. Rather, he should get a spirited little dog that will display his authority by demanding treats that will harm no one but him. Nevertheless, other animals, especially the cat, are preferable as pets for older people.

An older woman is walking down the street and hurries up to a huge dog. "Oh, what a lovely doggy!" she exclaims, her head lowered affectionately.

The "lovely doggy" gives her a nip and the master is berated: "A dangerous dog ought to be muzzled! And how do I know that he doesn't have rabies?" Of course the dog has been vaccinated, or else....

The animal has not done anything wrong. The woman was entirely at fault; she had only gotten what she deserved. Naturally myopic, the dog must have noticed something throw itself upon him—something, moreover, with its head bowed as if to bite him. So he bit it first.

Colette always recommended extending one's hand to strange dogs before petting them. She was perfectly right; they are olfactory animals. One should then remain still so the animal can have time to assure himself of the good intentions of this animal standing on its two hind legs.

The upright position is an attack stance for animals, and also one of dominance. This is why one should never get down on all fours to play with a dog, especially a large one.

From the age of six weeks to about eight years, a dog is learning. Thereafter his experience is fixed and he can no longer be changed.

Like any other animal including man, he learns in two ways: by himself and by being taught, first by his mother and then his master. His mother growls to make him behave; this growling is what the owner's commands replace. But the dog is a born observer; he is primarily responsible for what he learns.

Look at him. Lying on his paws, his ears erect, his nose quivering slightly, his eyes bright: Everything about him suggests the liveliest curiosity. He studies his new family. Soon he will know it by heart, down to each member's slightest gesture.

And he will be conditioned completely in relation to his family. However, unlike the cat, his conditioning will be more Pavlovian than experimental. The sound of the elevator means his master is coming home, the doorbell announces a visitor, etc. This kind of conditioning can make him appear silly: How many times does it happen that a dog will bark even when he sees you ring the bell and therefore knows no one else is coming?

A dog will not ask to go outside when his master does not have his trousers on. In bohemian families he learns very quickly that there is no special feeding time. I know one dog who doesn't lift a hair when his mistress puts on a fur coat: He learned long ago that she never takes him for a walk when she is dressed that way.

Since we have learned the dog's language, we should, in order to understand him, use it to the fullest. For his part, he will learn between 30 and 100 words, sometimes more, depending on his breed. Similarly, he will remember our vocal intonations: tenderness, defense, anger, order....

We have seen previously that punishment and especially reward play a big role in his training. The mere presence of a person and the caress of his hand are a reward. But a reward can take symbolic or even ritual form. Thus Youka, when she completed a film or posed for photos, had her fee paid in little dog biscuits, which she came to look for right after she heard the click of the camera and which she ate greedily—even though she hated them ordinarily!

Most animals can regress. A herbivore will find a wild meadow to replace his pasture. A bird will catch the first insect that comes along. An abandoned cat, too, will change into a wild feline and, if he is not too old, will learn to hunt.

This is true of all animals ... except the dog. He alone cannot fully retrace his steps. When, a long time ago, he was called "Tomarctus," his prehistoric name, he was already conditioned by man!

THIRTEEN
The Divine Cat

The cat was honored as a divinity in Egypt. Is it because of this past that in Europe in the Middle Ages the cat (especially the black cat) evoked an image of fire and brimstone?

In Paris and subsequently throughout France, he served as a scapegoat. Boys and girls threw him in the fire as an offering to the Lord, à la St. John. Then, joyful at having burned their demon, they danced around the expiatory funeral pyre.

Fortunately, Perrault came along, as did the Countess of Aulnoy. *Puss in Boots* and *White Cat* were destined to enchant people and reconcile them with felines.

England was less cruel; as early as the fourteenth century she rendered justice to Raminagrobis.

Did not the famous Whittington, elected lord-mayor of London three times in the fourteenth century, owe his good fortune to his cat? History, even if it has become legend, tells of this extraordinary side of the little feline, half-fairy and half-witch, bearer of good or evil.

Dick Whittington was still a poor clerk when his employer, a progressive shipowner and socialist, invited his employees to bring any objects which they had to sell abroad onto his ship; they would be paid for them. But Dick was so poor that he had nothing to offer except his cat!

A cat good only for killing rats; of what benefit was that? A great deal! At that time the palaces of the king of Morocco were being destroyed by rats and he could no longer even dine in peace since this horde was shamelessly coming right up onto his table to eat. In desperation he promised half of his fortune to whomever would rid him of these monsters.

The ship captain came to the king with Dick's cat, who proceeded to fare better than any courtier had. Chasing the rodents out, he promptly rid the palace of them entirely! True to his word, the king paid Dick in gold dust and diamonds: ten times his weight!

And thus was Whittington's destiny changed.

From Poland to Ireland, and even beyond, as far as Persia and distant China, cats—good or evil—haunt legends. Why? Why this ancient cult, these witches' dances, these marvelous stories, these cat-women? No one can really explain it. This is the mystery of the cat. Lewis Carroll was most misleading to reduce him to an enigmatic smile.

Perhaps the origin of these fantasies is the cat's astonishing vision, which allows him actually to see at night: Is it not diabolical to be able to see in darkness? (Furthermore, it has almost been established—scientifically—that the cat "hears" with his eyes!) And what about those inexplicable antennae— are they whiskers or eyebrows? At least the good, trusty dog doesn't have this disquieting feature.

However, the cat—that outcast—is an ideal companion for the intellectual, the hermit, and the elderly. He is less appropriate for a small child; the cat likes to maintain his dignity and resents the child's often brutal mischief. Actually, the cat is a homebody that likes everything just so. He has no taste for travel, likes only quiet people, and, if he does accept the presence of another cat, does not live with him as much as alongside him.

Celibate, individualistic, and egotistical like the leopard and tiger which he is in miniature (the lion being, as we have seen, the only feline to have a family organization), the cat concedes to a human being only what he wants to concede.

Therefore, man must accept the cat's behavior and must not impose his own on the cat. A status quo must exist between man and cat.

In a story that describes this animal perfectly, Kipling had him say, "I am not a friend and I am not a servant. I am the cat, who goes away by myself and for whom all places are the same."

And the woman who was victimized by his tricks observed, "O Cat, you are as clever as a man!"

Since then, modern scientists have confirmed what the poet divined: Of all the animals, the cat's intelligence and character are closest to human behavior. If you want to understand him and to be understood by him, you should never forget this.

However, Kipling was wrong when he stated that the cat is not "the friend of man." To be sure, he is not the companion of man the way the dog is, but he is capable of choosing a person voluntarily as a friend and even, in certain cases, of loving him deeply.

It is very possible that a cat will not reach an understanding with his master even if the cat has nothing for which to reproach him. If so, one morning, be he a farm cat or a city cat, he will disappear. All roads are inviting to him, whether they go past fields or chimneys: "All places are the same."

And, even if he is perfectly in agreement with the person with whom he is sharing his life, it can happen that he will be smitten by someone else and will leave his owner for his new love. It is best not to reclaim the cat by force: He will never forgive this.

Dove, a ravishing Siamese, led a happy life in the country—a perfectly orderly life, just as felines like. Every day she disappeared to have her siesta between 1 and 5 P.M.; the old woman with whom Dove lived knew that it was useless to call her then. Then, at 5:05 sharp, Dove returned and rubbed herself gently against the legs of her friend.

One day Dove failed to perform this ritual. Her mistress

waited until 5:30 and then became nervous; the woman searched everywhere, calling Dove's name. Her search took her to a campground about a half-mile from her home. Of course, the campers said, they knew the pretty Siamese who had her daily siesta in the Morins's trailer. The Morins? They had left today; the little cat had arrived just as they were taking off and had jumped into their car. "The Morins were very touched." So, convinced that the animal had been abandoned, they had taken her with them.

The woman was very upset. "But we understood each other so well!" Oh, yes. But how can one resist a new infatuation? Dove, despite the affection in which she held her old friend, had chosen those whom she now loved.

This is neither an anthropomorphic interpretation or an isolated case. Thousands of similar instances could be cited. The cat puts love before everything else. He can love one person exclusively and still flirt with others. He adores receiving visitors; he eagerly meows at them appealingly and asks them to pet him, at the same time glancing at whoever is his true love—it is fun to evoke jealousy! Doesn't the cat have a "human nature"?

One cat was seized with a passion for a friend of his mistress. As soon as the friend arrived, the cat made her feel welcome. When the visitor left, the cat cried so much that his owner, heartbroken but understanding, thought about giving him up to the "other woman."

Then one day the friend asked the woman to lend her her apartment so that the friend could entertain her illicit lover there. (Close friends do not refuse to perform such a service.) In this way the cat learned that he had been defeated by a human rival. When the cat next saw the woman who had "jilted" him, he ran away and hid under a piece of furniture, and refused in no uncertain terms to be petted by the hand that he had loved so. He never forgave her and, repentant, returned to his mistress.

For the cat that loves passionately, "breaking up" can be as tragic as for a human being. It may give rise to despair that borders on neurosis.

This was the case with a five-year-old black Persian when its elderly mistress died. Taken in by the daughter, the cat found as much affection in her new home as she had received in the old one. But it left her cold. She rolled herself up in a ball in a corner and never left there. She refused everything that was offered to her: both affection and food. She ate only one or two mouthfuls from time to time, just enough to keep herself from dying of starvation. Then she would go back to sleep as though life no longer held any interest for her. She no longer even performed her toilet, which convinced the veterinarian who was called in on the case that she had a genuine, serious nervous depression: A cat that no longer cleans itself is very close to death.

This state of lethargy lasted two years. Two years! Few humans would be capable of giving a comparable demonstration of their devotion.

But the Persian had lived in the country before coming to the city, and this is what saved her. One day she was taken back there. Unable to resist the sight of a garden, which reminded her of the old days, she broke out of her inertia for the first time and went for a walk. She was left in the country the whole winter, sometimes all by herself—which didn't seem to matter to her. Then one morning she was seen to pass her pink tongue carefully over one of her paws and to wash her muzzle. Her cure had begun.

The cat's territory is so essential for him that ordinarily he assigns it more importance than he gives his master. (By contrast, the dog puts more value on his master than on his territory.) Accounts of cats who flee from a new home in order to return to the old one are innumerable.

A couple lived in a suburban home from which they were

evicted to make room for a highway. Relocated in the city, one evening they noticed that their tomcat had disappeared. After several days a shopkeeper in their old neighborhood notified them of the cat's presence. He had been found at the site of their old house. It was really at the "site," because the bulldozers had destroyed it and had leveled the land; nothing remained of it. The cat didn't care; he had found his old territory and had set himself up in it again!

This story illustrates another feline enigma: His bewildering sense of direction.

Doctor Méry relates that, when he was a student in Marseilles, a museum guard decided one day to get rid of his cat, which was too old for his liking. He put it in a sack and threw the poor creature into the sea, several miles from shore. A month and a half later, the man had the greatest fright of his life: The ghost of the cat that he had so villainously assassinated appeared before him! The ghost? Not at all! It was the cat in the flesh that had returned to him.

That the cat had extricated itself from the sack with the aid of its claws and teeth was not so astonishing. That it had swum to shore is even less remarkable, since cats are excellent swimmers, contrary to what people believe. But that it had crossed a large city and had succeeded in finding the museum where he lived—that was very astonishing. Doctor Méry explained this exploit in a way that he himself said should be taken for what it was worth: The building was adorned with a large clock which sounded the hours and which could be heard from a distance. Given the excellent hearing of cats, it is probable that this one was guided in finding his way by the sound of the clock.

But there is not always a sound to guide those cats who travel as many as 200 or 300 miles to get back home. It is believed that this is a matter of interpreting sounds and odors but also, and perhaps most importantly, of observation. Lastly, at what point does parapsychology enter in? But here one must tread very carefully.

The dog is civilized to the extent that man is. But the cat can join civilization and adapt to it, while at the same time preserving the independence of a wild animal.

This is why he is untrainable. Actually, only an animal with social responses can be trained. The lion, for example, can be trained much more readily than the leopard; the latter 100-pound beast proves to be four times more dangerous than the king of beasts, who weighs 600 pounds! This also explains why man—contrary to what he can do with the large felines—cannot train a cat: The smaller the animal's size, actually, the more individualistic he is and hence the more difficult to train.

Therefore the owner of a cat must hold reasonable expectations. He will never be his leader except relatively. And, aside from propriety, which is natural in this feline, he will never get much out of the animal. One has little hold on this "gourmet," after all, if the animal can go almost entirely without food, even if he also appreciates what he is given to the extent of developing perverted tastes—like man!

Thus, I know a little cat who loves hors d'oeuvres and has a marked predilection for those with onions and pepper! Another, more classic, case loves bread, preferably black bread, followed by rye; white bread comes in last.

The dog is a glutton, but a voracious cat is abnormal. Like people, he "compensates" if he lacks affection, if he feels alone, if his territory is changed, or if his sexuality is artificially suppressed.

The cat is not an obedient animal. But his power of observation is remarkable. This has no relationship to how he has been brought up, but to what is referred to scientifically as "insight": the instinctive discovery of a solution. This form of higher mental functioning is also valuable to him in his relations with man.

Thus, one cat observed how his master opened the door by turning the handle. Later the cat climbed onto a table to get near the handle, and did just what the human had done. When the handles were replaced by doorknobs, which he could not take between his paws, he almost had a nervous breakdown. What annoyed him was not being able to do what the man could. He wasn't interested in going outside, but in opening the door!

The cat is a tease. He loves to play tricks, just like a child. He hides under the bedspread, being careful that not even a single hair is showing, and waits for the hand to draw back the spread. Then he rolls around with joy, just like a baby shrieking with laughter. If somebody comes just as he is trying to hide, he is frustrated and furious and he stalks away.

He has a wild animal's powerful sense of dominance. This also makes him arboreal. Perched in a tree, the cat dominates, sees what is coming, does not run the risk of being surprised by anything larger than he, and leaps on his prey before it has a chance to flee. Carrying his wild habits into civilization, he loves the tops of armoires and readily sets up his keep there. For the same reason, he is displeased with modern apartments where objects are likely to be too close to the floor. He feels defenseless down there, at the mercy of whoever comes along, resentful and uncomfortable. Again contrary to the dog, he does not allow himself to be "taken advantage of." A cat will readily jump onto his master's lap but if put there forcibly he will not stay. If the person tries to make him stay, he will quickly turn aggressive.

The cat has typical feline vivaciousness. But he is also possessed of an extraordinary optical instrument: his eye. With his eyes reduced to narrow slits, he pretends to sleep whereas in reality he takes in the dimensions of the area and at one and the same time sees the hand (which thinks it will capture him) and the place where he will hide before being caught. For he comes and goes at his own pleasure, not at man's.

Nevertheless, this capricious creature is extremely reasonable. Very hardy and a great sleeper, he does not like to extend himself—except to play (he remains playful to the end of his life). So he needs only a little space—a single room suffices—and little food.

Accidentally imprisoned in a tool shed by his masters, a cat stayed three weeks without eating or drinking. He was found there alive and simply a little ... dried out. Any dog would be dead: he would have killed himself by barking or by his vain efforts to escape. By contrast, this feline remained quiescent, compensating with sleep for the lack of food and not exhausting himself in futile efforts. That is what saved him. Similarly, sleep—the cat's weapon—may have been what saved a young man who was trapped under ruins in a Rumanian earthquake for 250 hours.

Cats have all the attributes and shortcomings of the feline family: pride, irritability (although readily angered, the cat accepts being punished if he deserves it and will not retain any rancor toward his human friend), independence, inability to be conditioned, and an unfailing sense of dignity. A cat who returns home at a gallop will stop before going inside, in order to catch his breath and, if his run has messed up his fur, to put everything in order with three licks of his tongue. His tenderness and his desire for and pleasure at petting attest to his great sensuality; all that is necessary to be convinced of this trait in him is to watch him roll around in the sun.

This is why it is extremely important to wait until a male is a good one year old before having him castrated. He will then be a "tomcat": Sure of himself, he will have reached temperamental maturity normally, which parallels sexual maturity. If cats are castrated too early, they will be imbalanced and will become aggressive—just like female cats who have had their ovaries removed too early.

These are the cat's general characteristics. Each breed has a different personality.

The Persian has the lion's mane and also his character. His sociability is very pronounced, as is his family-mindedness: the Persian loves being "married." Confident of his strength, he can impose it on others, lay down the law, and exact respect. Woe to the alleycat who, proud of his thousand scraps, rebels: The Persian will give him a drubbing which will put him in his place. Short-haired domestic cats know this well and avoid quarrels with this imposing-looking cat.

Like the Yorkshire terrier, the handsomest of our deluxe cats has a proletarian origin due to utilitarian reasons. At the beginning of the nineteenth century, the warehouses of Liverpool were invaded by so many rats that the products they stored—among others, magnificent rugs imported from Persia—were seriously threatened. People tried everything to stop this destruction but their cats were killed by rodents stronger and more powerful than themselves, and even the canine ratters left some flesh behind.

This took on the proportions of a national catastrophe: for their daily lunch, the rats were gnawing England's treasure to pieces! So the city of Liverpool promised a reward to whoever could "make" a strong, robust cat whose fur was so thick that rats' teeth could not reach the skin, and whose nose and ears were so small that the rodents could not get hold of them.

The Angora, a long-haired cat, already possessed some of the desired characteristics. Starting with this cat, a strain was developed with thick fur, miniscule nose and ears, and powerful muscles. The resulting cats were then released in the warehouses and emerged as the victors in this merciless war, thus saving the Persian rugs. And this is why they are called Persians.

The Abyssinian, a descendant of the Egyptian cat, may

have been the first cat on earth for just that reason ... unless the first cat was the domestic short hair. People tend to scorn him, treating him as an "alleycat." They also tend to regard him wrongly as a nitwit (which no cat is), even though he is a very intelligent animal with every right to be proud of his ancestry.

Nevertheless, certain genetic characteristics of the domestic short hair are astonishing. Thus, a white cat with blue eyes is almost always deaf; a tricolored cat is usually female and if male is sterile and sometimes even lacks testicles; a ginger cat is usually a thief; and as for the tailless Manx cat, he is one of the mysteries of genetics, too.

The origin of the Siamese, a genuine little wild animal, near relative of the wild cat, has been much debated. One genetic theory holds that the velvet gauntlets that he wears so elegantly were originally the ocellate spots of a wild animal, *Felix bengalensis*, whose spots had "fallen down" to his paws. Moreover, the Siamese has the proud and ferocious nature of the wild cat. Intelligent, mistrustful, and with lightning reflexes, he can love his human friend deeply but be quick with his claws toward other people. Of all cats, his habits are nearest those of arboreal wild cats.

The Angora, Abyssinian, domestic short hair, and Siamese are the four great feline breeds, from which stem the Khmers, Carthusians, Burmese ... Unlike the dog (ranging from the one pound Chihuahua to the German mastiff or St. Bernard weighing a good 175 pounds), the cat never exceeds its normal range. If it does, it is called leopard, ocelot, or tiger!

A lot of people say that the cat is "incomprehensible." This is false. To understand him, anthropomorphism is useful. We can understand his character through our own, remembering that, like ourselves, he is perfectly capable of higher order responses. At least, our understanding of him leads us to believe this.

If the dog and the cat knew how we regarded them, they might be quite surprised. And how do they regard us?

FOURTEEN
Man and His Dog

The dog is not more intelligent than the cat. But his character, his sense of things, is different. A general can be a genius, as can a poet. Let us say that the dog would make a better soldier, and that the cat could be a sublime writer.

The dog can be trained, the cat can not. This does not mean that the cat lacks intelligence. If he wanted to, he could be trained; he does not want to. This is not a proof of his intelligence either, it is simply a sign of the cat's independence. The cat is the little cousin of the jaguar, the dog of the wolf. The large cats do not obey any law, whereas the wolf respects that of his group.

In order to get along well with a cat, one must not ask anything of him. And if you want your dog to be happy and perfectly comfortable with himself and with you, he must be trained.

I recently met a nice hippie who was fondling a puppy. He candidly explained to me, with a big sigh of resignation, that when the pup became mature he would release him in the woods in order to let him return to his normal life, the true life of the wild, free dog. In his ingenuous anarchism, our friend was very surprised to learn that the dog was absolutely dependent on a human companion and that, left to his own devices, he would die. He was surprised but very happy to know that he could keep his dog without violating his principles.

People tend to belong to one of two opposing groups vis-à-vis animals: those who want their dog to snap to attention, and those who let him urinate on their friends' rugs because they do not want to bully him: "I abhor slavish dogs!" But a dog whom one trains—trains well, because of his language and his habits—is no more a slave than the child that one forces to go to school. In fact, often the dog is more interested than the child!

In the first category are often found, unfortunately, the lovers of wolf dogs. These people are aggressive types who, whip in hand, can bellow like drillmasters. They delight in having ferocious animals with whom to threaten the whole world. The German shepherd is actually the most popular dog today: one indication of the aggressiveness of modern man. In 1920, at the end of the war, the Pekinese was preferred.

Essentially, the German shepherd is just a gentle dog who insists that his rules be obeyed (this is why he is made a watchdog). His bad temper is not innate but acquired. For this state of affairs people are usually solely responsible—people who do not hesitate to use Gestapo methods: intensive night work (when the dog wants to sleep and therefore is in a bad mood) under the light of special lamps and always with the whip. This is not training; it makes the animal insane—as insane as his master! To act this way is to flout the moral standards of man and beast.

These same animals can be made into guide dogs for the blind: the calmest, most amicable, of dogs. They can also be made into capable mountain dogs, better even than the famous St. Bernard at rescuing hapless Alpinists. Without any doubt, they represent the most intelligent of canine breeds. And man has certainly taken advantage of this intelligence: There have been paratroop dogs, police dogs, detective dogs. . . .

The Paris prefecture has a school for dogs that is a miniature Ecole Polytechnique for German shepherds. Not just anyone, whether master or student, can get in. Participation is an honor for both parties.

I have often been present at these "lessons." Systematically perhaps, but with an infallible "instinct," man teaches beast by relying on genetically based rules that have ordered the pack since prehistoric times. But the trainer never—and this is an absolute law there—resorts to a whip in the course of the work. Furthermore, he has no need whatsoever to use one. Having recognized him as his leader, the dog obeys him implicitly.

"Ah," beams the taxi driver, "if only you could see my poodle. A real clown! He loves to show off: You put a piece of sugar on his nose, and he jumps up in the air and catches it in his mouth before he lands on his feet! He never misses!"

The customer is either delighted with the driver, or irritated. Is a dog supposed to be a clown? Fooey! Is this one likely to have come from a long line of circus dogs so that this is what he needs to make him completely happy? Often it is the dog himself who, by some improvised stunt, lets his master know that he can teach the dog some tricks. These tricks amuse the dog; and the more he is admired the happier he is.

All animals, without exception, love applause. As silly as what they have been taught may be, it is much better than not letting them exercise their brains at all from the time they wake up in the morning to their evening feeding. The wild animals to be pitied the most are not those who work with a trainer, but those who spend their lives in cages without doing anything.

Without doubt, the poodle has the soul of an actor: He asks only that he be allowed to show off. But it would be unforgivable to have a wolf dog or a mastiff perform the same tricks and to make fun of him.

This is why, during training, one must always respect not only the dog's individuality but also the inborn characteristics of his breed. A watchdog can hunt on his own but this does not make him a hunting dog. And if a hunting dog barks at an undesirable visitor, this does not make him a watchdog.

The watchdog observes the basic rules that regulate the pack in order to protect it against intruders, in this case the human family being regarded as his pack. In most cases a true watchdog has inherited these rules. All that man has to do is to bring them out.

The dog barks. This is a vocal warning: The intruder must not enter. But if the leader—man, under the circumstances—lets someone enter, that is all right. On the other hand, the stranger who enters finds himself effectively incorporated into the pack, and if he wants to leave he will be prevented by the dog—as a shepherd dog does with a sheep or cow that strays. Only the master can conduct the stranger outside the "pack."

Aside from these general principles, there are others which are extremely complicated. For example, access to the territory does not imply access to the core area or even to the whole territory. Furthermore, one must not take anything in the territory, for the notion of theft is solidly established in the genetic memory of the dog.

Uépo, of whom we have already spoken, was trained by a dog trainer who refused to teach him to guard. "It's innate," he declared admiringly. "I can't teach him anything. I would only spoil him."

He barks ferociously (this very gentle dog could certainly break a burglar in two) whenever a visitor comes. If his master opens the door and lets the person in, Uépo sniffs the newcomer for a long time (the guest is advised not to budge). Then Uépo wags his tail as a sign of welcome and allows himself to be petted. But he follows the guest into the room and unless the "pack leader" accompanies him, Uépo will not let him leave.

Thus, some workmen came to work in the kitchen. Uépo amiably came to say hello each day that they were there. His master had warned the workmen: "If you want to leave, call me."

However, one apprentice had not taken this warning to heart and, just as he was about to pass through the doorway, the dog gently barred his way. The apprentice wanted to go outside. So the animal got up on his hind legs, placed his paws on the man's shoulders, and pushed him back gently, without any malice.

"The second warning," said the master. "If you get bitten, it will by your fault." He was right, and the dog was, too. The young man did not argue.

But the funniest story of all took place at their country home. Uépo's masters went out one day, and as usual left their dog with the housekeeper. Upon returning that evening, they discovered the woman was still there; she was supposed to have gone home a long time before. Uépo had absolutely forbidden her to leave in his leader's absence!

"I tried all the exits," she explained. "Nothing worked. He didn't growl but he stood in front of me ... so I preferred to wait for you!"

In the pack, it is not permissible for a stranger or even a pack member to grab hold of the prey and go off to eat it all by himself. In the same way Uépo would not allow anyone to leave his territory. The result was that every morning a polite but definite misunderstanding took place between him and the housekeeper whenever she wanted to take out the garbage!

This woman was allowed to touch anything in the house except for personal effects that were in her employer's room (the "keep"). Without growling or showing the least ill will, the dog picked up a vest in his mouth that the housekeeper was about to put away and forced her to leave it where it was!

One day at a hotel, seeing the porter taking down the suitcases, Uépo hurried over and snatched them from his hands (the man was perfectly willing to release them). Then, satisfied, sitting next to this property which no one would think of taking from him, he waited for his master.

This was truly a case of canine behavior in the pack being transferred to the human family. One must always build on this base in training the animal. And, if one wishes to preserve the dog's qualities as a guard, one must respect these fundamental principles in his own actions—at the risk of appearing ridiculous to other people! If the dog is encouraged to be friendly immediately with a visitor—or if the latter has the right to leave without being accompanied by the master—the dog is not to blame when one day he benignly lets a burglar enter and leave. He will only be respecting the rules and instructions of his master.

With a hunting dog just as with a watchdog, one must make use of his past, that is, his genetic memory. Evidence of this is readily available: In wolf or dog packs, there are always those individuals who attack the game and those who stalk it. Among domestic dogs, there are pointers and retrievers. But a problem seems to arise here: how to teach the animal that is tracking the rabbit not to keep it for himself.

This is less of a problem than a person would believe who was unfamiliar with the ancestral ways of canines, for the hunt is a pack affair. It is a cooperative endeavor which involves almost communal thinking. Dogs hunting in a pack communicate with each other and give each other instructions.

It is useful to remember, especially at the time of training, that no wild animal hunts "naturally," but rather imitates its father or mother. Initially, hunting does not interest the young animal as long as his parents are feeding him. But after several months he develops a desire to imitate his parents and to play his proper role. But the parent will never force the young animal to hunt; he will only compel him to watch.

Man must respect this behavior. The best pack leader is an old hand, who will teach a young dog how to hunt much better than a human trainer could. For experience is passed on step by step: A child learns simple addition before attacking higher mathematics. We must demand no more of an animal. The principle to remember, perhaps, is, as in all matters, not to make any mistakes.

Thus, one must remember that a dog who has a disagreeable experience holds the place where it happened responsible. Nestor, four years old, is a French pointer from a good family. He knows his trade perfectly. But this was not the case with the friend who accompanied his master on the hunt one morning.

The dog scrutinized the friend out of the corner of his eye. The man had an unusual manner of holding his gun; it was always pointed where it shouldn't be, either at a tree, at his companion, or at Nestor himself—in which case Nestor wisely stepped out of the way: One never knows what is going to happen. Hunting dogs are horrified if they see the muzzle of a gun pointed toward them. On the basis of experience they know that this is dangerous.

Suddenly a hare made Nestor forget his apprehensions. He hurled himself onto the fresh trail where, a few seconds later, a volley of lead in his rear end made him stop! The friend, of course, had mistaken the dog's rump for the hare.

Two weeks later Nestor, the lead having been removed, had recovered from his wound. He again set out for the hunt, this time with only his master. The man headed toward the woods where he usually hunted and where the accident had taken place. But the pointer stopped in his tracks and refused to enter the woods. Nothing worked on the dog: neither scolding, nor petting, nor commands. The master was forced to change his hunting grounds.

Some Americans had the same experience with a dog that found himself face to face with a bear. He fought the bear, which his master eventually shot. Afterward, he always refused to pass by the spot where he had met the bear.

Watchdogs, hunting dogs, shepherd dogs have been used for a long time. Then along came the dog without purpose, called the "luxury dog" by the tax collector. This type of dog often belonged to an older lady who would go without steak to feed it to the only living thing she had left to love.

One cannot ask this dog for the services that one has a right to expect from other dogs. The various breeds that have been combined to create him have thoroughly blurred his genetic memory. It is just as though his computer program is scrambled. However, no matter how miniaturized he is, there is one thing that he will always retain: his sense of leadership and obedience to the leader. To this extent, the tiniest of Yorkshires is trainable.

The first thing he should be taught, as with all dogs, is "yes" and "no." These words correspond to two maternal growls, very different from each other, which distinguish what is permitted from what is not.

A badly brought up dog is disagreeable. The owner, like the father of a child, is within his rights to permit this but he does not have the right to impose the dog on others. These are the animals who steal the supper, urinate on the sofa, defecate on the sidewalk, and bark without cease. These are the scourge of all dogkind.

You must be thoroughly convinced of one thing: Even though they annoy everyone, they are not any happier because of it!

FIFTEEN
Turtles, Fish, Birds, & Co.

If for most people to "have a pet" means a dog or a cat, others prefer a bird, some fish, a hamster, or a turtle. Literally millions of fish and turtles attest to this!

This does not always reflect merely the desire to have these particular animals, it also may indicate a preference for ... silence. A cat meows, a dog barks, a bird sings. ... Amid all the noise of modern life, these animal sounds either are not noticed or they superimpose themselves on these other noises and are insufferable. It is a matter of personality: People who want to murder their neighbor because they hear him stamp his shoes on the ground and clank the elevator door while whistling generally prefer a mute animal.

For this avalanche of chelonians (the turtle belongs to the order *Chelonia*) and aquaria, the child too has a responsibility to bear: Parents who want to please a child who clamors incessantly for a pet, but who understandably do not want to be burdened excessively, readily offer a turtle or hamster. These animals, incidentally, offer another big advantage: neither one has to be walked.

The hamster is darling. His habits are the most amusing of all the rodents. However, he presents the major problem of dying young, at two years, or three at the most. This feature may be the cause of a great disappointment that the child may not take well, especially if he is between 5 and 15 years old.

Obviously, one does not run this risk with the turtle: He intends to live to be a hundred!

Whether they weigh an ounce or 1,000 pounds, whether they are marine, amphibious, or exclusively terrestrial, all turtles have the same physiological and biological characteristics. More modern than the snail (another animal in a shell), the turtle nevertheless can draw on a not inconsiderable past of 220 million years. Doubtless feeling comfortable in his carapace, he has not evolved over this time.

He is a cold-blooded animal whose temperature varies according to his surroundings. He is an animal of simple pleasures—merely a long winter siesta beginning in autumn and ending in spring. If he decides that the weather is too chilly then, after a light snack, he will resume his nap and await better days. In the wild state, he will dig a hole as soon as the cold weather comes, and will remain down there to sleep where it is warm. Transferred to the world of people, the turtle manages very well all by himself if he has a yard. But in an apartment he needs to have his natural habitat recreated. A box full of soil and dead leaves will furnish the nest that he needs to hibernate. He will leave the nest by himself when he has decided to do so. Man should never intervene: The turtle alone knows, biologically, the correct length of his sleep.

The amphibious turtle is especially interesting. Since he needs both land and water to survive, he should be given both when he joins a human family. If he is tiny, he will live in an "aquaterrarium" combining water and land. If larger, he will adjust to a yard and a little pool, as long as the latter's sides have a gentle enough slope to allow him to climb out. This very old boy is no acrobat and if he has to make dangerous jumps to emerge from the water, he runs a good risk of falling on his back—a fatal position. In this environment he will choose his lunch himself, just as he would in his natal pond: beetles, larvae, insects, etc. Many people do not know that the amphibious turtle lives in his natural state in many places; the ocean is not necessary.

There is but one precaution to take: Prevent a sudden cold snap from surprising him in the water before he has had time to go hibernate in the little mound of soil, dung, and leaves which has been provided for his winter comfort.

We had a case of this with a turtle who, in the course of winning a race with a hare, strayed onto a highway. Brought back home safely to his yard and basin, he was leading a happy life when winter arrived early. The water froze. Our turtle was caught in a tomb of ice. But a little thing like this did not discourage him. With his beak and his claws he broke through an inch of ice in order to breathe.

Whether terrestrial or aquatic, this little monster—a throwback to prehistoric times—is readily tamed. He responds to the sound of his name and, just like a cat, loves having his neck scratched! If a person has the misfortune of displeasing him, he will see to it that the person "feels" bad about it at once: He will direct a particularly bad smelling jet of urine at whoever is holding him against his will!

The fish, symbol of the early Christians, is the symbol of luck to the Chinese: The year of the fish is anticipated with joy, for it will be propitious.

This cold-blooded animal often has "warm" behavior: He is much more affectionate than he is believed to be. Once his usual world is recreated artificially, he is perfectly capable of showing friendship toward man. All aquarium lovers agree: When they return home, the common goldfish or the sublime fighting fish make merry! In their own way, of course, for if they warble soft hellos, man's tin ear, incapable as it is of perceiving ultrasound, cannot hear them. One thing is certain: The little fish, if left all alone in his aquarium, will not grow—he will let himself die of melancholy. If his owner returns only when the fish is asleep, he will waste away very rapidly.

We have already seen that fish are endowed with an intelligence and memory (often biological) that are poorly understood but that we are sure exist. If we want these animals to get along well with "their humans," we have to understand them and be as familiar as possible with the habits and behaviors of the species that we have.

Thus, one quickly learns that the scoop with which he intends to seize a particular fish does not frighten the other fish at all. Their curiosity, which is similar to man's, is such that any unusual object that plunges into their habitat has the effect of attracting them: they surround it.

The external world also constitutes a spectacle that fascinates them. If we like to watch fish behind the glass that separates us from them, they enjoy observing us with no less interest!

But what the fish enthusiast must know—at least if he does not like melodrama—is that the sympathies and antipathies of this world are without nuances. Fish can nourish a hatred that can be deadly. The guppy family gets along "swimmingly" with the xiphias. But just invite a gudgeon into their home and they will kill him in no time.

We do not understand the aquatic world very well, especially since our senses, which are very different, allow us almost no communication with it. Nevertheless it captivates us. Its attraction for man dates from far-off antiquity and has been exerted on all peoples. The ancient Chinese, lovers of monsters, succeeded in fabricating some extraordinary hybrids.

These curious animals offer more to us than their variety of appearance. If we are truly interested in them, what they reveal to us is exciting. Further, fish enthusiasts are increasingly fanatical. As with bird lovers, this enthusiasm can almost become psychotic: Psychiatrists haven't seen the end of "zoomania" cases.

Given the amount of interest that they generate, these animals are in a position to insist that whoever takes care of them not rely exclusively on electrical pumps. Otherwise they will hold back plenty of surprises. Thus, like large animals who readily kill their young in captivity, aquarium fish—even the least carnivorous of them—are quite likely to kill their fry. This is why fish enthusiasts often have a second, somewhat smaller aquarium which serves as either a nursery for babies wrested from the gluttony of their ogreish parents ... or a hospital.

It should also be remembered that, just like man, the fish is not monolithic but variegated. He can be jealous, or faithful ... Don Juans are even encountered, and are not always male: In certain species the females demand several lovers. And in some species the males eat their milt, either because they are simply gourmands or because they have heard discussed the woes of having too many offspring. All these "faults" are magnified by captivity.

Depending on their tendency (and the size of their habitat), most aquarium fish range from 1 to 16 inches in length. Of the 40 million species of fish, about 2,000 can survive in captivity.

Fortunately for the fish, modern aquaria are no longer the goldfish bowls which graced the Utrecht-velvet salons of our great-grandmothers. Modern tanks, with their aquatic vegetation, even try to suggest the deep: We can only hope that their inhabitants feel just like fish in water.

Nowadays Little Miss Muffet is called Cathy; she no longer sits on a tuffet and embroiders. She is a model, a secretary, or a stewardess. But she always has a canary at her window.

Formerly, when she spent the day sewing near the cage, she "chatted" with her feathered captive. And when she shed tears over her unhappy romances, the trilling of the joyful canary consoled her.

Nowadays, since her mistress does not work at home, the little bird is all alone. Having no one to charm with her singing, she keeps silent and eventually dies of boredom ... like a fish! For these winged creatures probably sustain isolation even worse than people. Only a leopard flees its kind!

As we know, birds are much more intelligent than is commonly believed. They know their master very well, and love to see him and "talk" to him. It has been observed that they adore music, especially that of Mozart. In the absence of a mate or a companion, therefore, let them have the pleasure of listening to a Mozart symphony or, better yet, Olivier Messiaen's *The Birds!*

On the other hand, they should not be placed in front of the television, even to entertain them. They are totally allergic to it. The little screen is capable of causing them to molt repeatedly. Why? No one knows yet. Doubtless science will eventually give us an answer, but for the time being we have to content ourselves with recognizing the phenomenon without explaining it.

On the other hand, it cannot be denied that certain bird lovers understand their behavior better than anyone. The most famous of them was the bird man of Alcatraz who specialized in canaries and was known to all bird fanciers. His story is quite extraordinary:

It begins like a western. At the end of the last century in a town in the American West, a 19-year-old boy accidentally killed a man in the course of a fight. He was sent to prison. He was never to get out. He married there and died an old man. But in the meantime he transformed his cell into an aviary and became a world-renowned ornithologist.

Amateurs like this have taught us a great deal about the cruelty of our charming companions, a cruelty doubtless exacerbated by being caged.

Several species of birds lived in an aviary. It was noticed that the eggs of a couple of widow birds were regularly broken. It took days and days of observation to discover the culprit guilty of this "infanticide."

It was a cardinal. The bird seemed to be on excellent terms with his neighbors. He never exchanged pecks or even threats with them. He disguised his hatred for the future progeny—or for their parents perhaps. He also exhibited remarkable patience, since sometimes he had to wait for several days before enacting his crime. The father and mother rarely abandoned the nest simultaneously. But as soon as this did occur, the cardinal took advantage of these several seconds of inattention by the parents and destroyed the future brood.

He was separated from the other birds. He was given a female. She laid. And he was a model father! Had he been jealous of the other birds' family life? Who knows what goes on in birds' heads?

The most fascinating birds, without any doubt, are "talking" birds. It is now known, contrary to what was formerly believed, that they do not understand the exact meaning of the words they pronounce but they almost always understand their general significance. Here again, the presence of man is important for them.

Living in the wild, the parrot is arboreal. Monogamous, he is extremely faithful to his companion: He defends her vigorously if he believes her threatened or if another male approaches her.

These characteristics are also found in his behavior toward man for, unlike most birds, the parrot is readily tamed. Then he is totally dependent on having a "family" and must be treated as though he were one of its members. He always has one favorite and is very jealous. He thrives on "sweet talk," having his neck scratched, going for a walk on a finger, and tidbits. He is only happy in a loving atmosphere. And he will learn on his own the words which correspond to a gesture or an act. The number of parrots that say "Hello" when the telephone rings or "Mama" when the mother of the family enters the room is incalculable. The most talkative are the Amazons, especially the Gabons, capable of learning a hundred words.

Unfortunately, the pleasure of having a talking bird is outweighed by the same odious exploitation that has released little wild animals on the market in recent years. Captured with nets, put into sacks, sped to their destination like any commodity, scarcely 30 percent of them will still be alive upon arrival. Those that are still living are sold right away.

The actor Francis Blanche had a mynah that, as soon as he saw his master in the distance, called out "Papa! Hello, Papa!" at the top of his lungs. He never made a mistake and called any other man "Papa."

The mynah, belonging to the same family as the starlings, is among the most interesting of birds. The best talker in the winged world (outside of the *Psittaciformes*) is doubtless the great Javanese mynah. One of them, Raffles, was an American star: He appeared at the Metropolitan Opera in New York where, not letting himself be the least intimidated by the thousands of spectators, he whistled the national anthem!

During the Second World War another mynah, whose "sailor's language" was full of tartness and eloquence, was assigned the duty of selling defense bonds, which he did to the tune of 15 million dollars!

But a mynah that is bored, is lonely, or does not get enough affection will refuse to say a word—just like a parrot.

The behavior of certain birds is so close to man's that it is easy to see why they readily live with and get along with us.

A young woman had a blackbird that lived with her for several weeks. One day he took advantage of an open door and flew away. But not very far: He only flew onto a wall about the height of a man. There, with his saucy black eyes, he defied his mistress.

She approached him cautiously, holding her hand closer and closer, and then snatched at . . . nothing but air. The bird just flew away. He perched on a balcony ten yards away and waited for his mistress. They repeated the same game.

This time the blackbird perched on a branch. As the young woman had lost sight of him and was looking for him, he whistled sweetly. Having thus assisted in the search for himself, he waited until she had gotten right up to him before taking off again. He continued to mock her, touring first the house and then the whole village in this way. And whenever she lost sight of him, he called her!

Night came. Everyone was asleep. The next morning the woman awakened with heavy heart, convinced that the bird—for he was, after all, only a bird—had left forever. Then, as she was making coffee, three mocking notes called her from the window sill!

The game lasted two days. Then, having teased the human creature sufficiently, the blackbird left for real. Doubtless if he had known his mistress for a longer time, he would have returned home one day, like Jacotte.

How can this behavior be explained without resorting to anthropomorphism? The "mocking bird" is not a myth. He makes fun of people who are not satisfied with the real companion that has been offered them, namely the dog, but have to imprison wild animals under the pretext of loving them.

But no bird resembles any other, neither in plumage, nor song, nor character. There is as much of a difference between a macaw and a canary as between a shark and a gudgeon. Some species are extremely sociable and seek out man's company. Others, no matter what you do, always remain indifferent. It is up to whoever likes them—that is, who likes them in cages—to choose.

But if he does not want to make his birds too unhappy, he should learn about their character and their behavior, and conform with them. If he has an aviary at his disposal, he should study the wide range of habits of the species which he confines there. Like fish, certain birds are xenophobic and only tolerate members of their own family. Others, more high-minded and conciliatory, get on well with some species but are still "racists" when it comes to living with others.

One might say that man has become "animalized." He has reached the point where, if he wishes to enjoy the nightingale's song during the day, he pokes its eyes out. But has the animal become "humanized"? Perhaps, if one believes the astonishing evidence of Professor Cunin.

The novelist André Demaison had two pet crows who had

again taken up their feral existence with the other crows. But they were so tame that when he went for a walk he would suddenly see the two cawing crows land on his shoulders. They would then come home with him—to "their house." They would leave after several days, then return....

The novelist died. At the instant when the funeral cortege left his house, a flock of crows arrived. Like a huge mourning veil, they stretched out high above the coffin and the procession of mourners. This was 15 years ago, in a rural area; the hearse was still pulled by a horse that trotted slowly, ceremoniously. It took a half hour to arrive at the church. All this time the crows accompanied the cortege, flying in silence. So strange was their presence that all the mourners walked with their eyes looking upward. And the birds did not disperse until the coffin had entered the church.

This is an anthropomorphic tale based on a coincidence. Scientifically there is nothing else involved. But just as scientifically, one thing is certain: birds communicate among themselves. In *What to Do Till the Veterinarian Comes* we told the story of the crows who drove an entire Irish village crazy. One of the birds had discovered that he could use his beak to open a bottle of milk that was left every morning in front of each house. Having developed a taste for this liquid, he informed his pals, who were eager to imitate him.

Of course, it should be quickly added that animals know nothing of death. But hard facts like this should not stop us from speculating. After all, as Jean Cocteau said, science contents itself with proving what poets have dreamed!

SIXTEEN
Man's Noblest Conquest

The mare was named Roxanne. She was a thoroughbred Anglo-Arabian. High-strung, haughty, and stubborn, she always went her own way.

One brisk autumn day she made so many mistakes on the racetrack that her angry rider punished her severely—which was wrong. Then, calmed down, he took several prudent steps backward in anticipation of the reaction, which, to his surprise, did not come. And so, convinced that he had been right, he went away satisfied.

Three days later, having forgotten the incident, the rider entered Roxanne's stall. He gave her a friendly pat and said cheerfully. "We're going for a walk, Sugar!" But before he had a chance to react, he found himself hoisted a yard above the ground. Roxanne had seized him by the top of his vest.

Vengefully, she casually let him fall . . . right into the feeding crib! He was imprisoned in this sort of cage. The horse watched him evilly, snapping at him menacingly each time that he made a movement. He was in a tight spot indeed. What could he do? He could only pray that someone would come to his aid. Someone did come, but only after several long, agonizing hours.

The horse is endowed with the same excellent memory as the mule and also has the latter's spitefulness: The vengeance of *The Pope's Mule* by Alphonse Daudet is not just a novelist's invention. Roxanne made her tormentor pay for the punishment which he had inflicted three days before. This is one of the faults of the horse—and not the least of them.

This quadruped is hardly what we think of as a domestic animal nowadays. However, for centuries he was man's boon companion; in part, man owed his own success to the horse. For a long period, only two ways were known for getting from place to place: by sea, the boat; and by land, the horse. The Egyptians and then the Cretans were familiar with the horse, and made use of him, well before the Romans.

But what prehistoric ancestor was the first to discover that this 600 pounds of meat had a sensitive spot, his mouth, which, when a bit was placed in it, allowed one to subdue the most recalcitrant animal? No one knows.

In the West, the horse did not assume his real importance until the Middle Ages, after shoeing, the horsecollar (which allowed him to be harnessed for draft), and the saddle had been invented. Before then, the rider rode bareback. Lazy kings had their carriages drawn by oxen not just because the slowness of these animals suited their slothfulness: Horses' hooves got worn down rapidly (despite the leather sandals which the Romans had invented for them), and this made long journeys impossible. Furthermore, the simple harness quickly started hurting the animal since it rubbed against his neck, thus preventing him from pulling very heavy loads (the Roman chariot was light, holding only one man who often used four horses to pull him). It took man until the thirteenth century to invent the simple horsecollar that draft horses have worn subsequently.

So "man's noblest conquest" became the workhorse, the drafthorse, or the riding horse. And just as there had been a Stone Age and an Iron Age, there came to be born an era of the horse, a new age of civilization.

As close as he became to man, the horse today has lost ground. Yet he still stirs up controversy, some people contending that he is the smartest animal on earth, and others—including scientists—holding that he is one of the dumbest!

The fact that he is a mathematical whiz does not, unfortunately, obviate the possibility of his stupidity. On the contrary, it may even constitute overwhelming evidence of it.

It is said that great experts at mental arithmetic are simple minded. This does hold for plenty of "prodigies," whose exploits belong more to the circus than to real science. In a sense, their talents seem very close to those of the famous Elberfeld horses. Just like these amazing horses, a German, Zacharias Dase, born in 1824, calculated the logarithms of the numbers from 1 to 100,500 in his head. However, he was always unable to explain what he was doing.

One of the most baffling things in the world is the explanations that these "mathematicians" give concerning how they get their answers. That is, the explanations they give when they give any at all: The majority of these people can provide no notion whatsoever, even the most rudimentary description. Professor Rémy Chauvin reported that when he talked with the mathematical whiz Lidoreau, the latter said, "I see the figures as apples in a tree. When I am asked a question, they move by themselves into the correct arrangement!"

But the relationship that exists between these "brains" and animals is more disturbing still to the normal human mind. It has been noticed that, like Amphion, these highly specialized people seem to attract fish and birds! Now, are not these animals also born mathematicians?

The horse is very sensitive. He is frightened by the least unusual object. This is one of the characteristics of his peculiar visual sense. He also is unusual in being poorly armed. Having neither horns nor claws, he has to rely on his speed to flee. He is quite fast but cannot run for any great length of time.

We also know that his size was originally that of a terrier. It is not impossible that subsequently he retained, genetically, the sense of inferiority which characterizes small animals. The horse has been tall for only a few millennia, which is quite recent. Witness Prjewalski's horse: Named after its discoverer, this little Mongolian horse is without any doubt a true descendant of the wild horses of old; and therefore similar to them.

Even though his training can be rapid and relatively easy, the horse is totally devoid of integrated behavior. Ignoring the warnings of Jesus, his right side is absolutely oblivious of what his left side is doing—even for the simplest things. Thus, if he is accustomed to being mounted on the right, he will never allow himself to be mounted on the left!

Therefore his conditioning is limited. His training consists mainly of letting him do what he does normally but getting him to do it in situations where he would not normally do it on his own.

His trainer must above all demonstrate great patience. He must teach the horse only one act at a time, in just the manner that will always be required of the animal. If not, the horse will get everything mixed up! This is the military mind par excellence: Attention! At ease!

If he does not understand right away, one must never get exasperated, raise one's voice, or show nervousness or brutality; to do any of these things would be the best way to get nothing out of him. For he does not understand why he is being hit; punishment only angers him. Unlike the dog, this "dullard" only understands rewards.

Actually, the training of a horse depends mainly on the trainer. The rider and his mount make a team, the perfection of which is accomplished only through perfect understanding. The slightest pressure of the rider's knee constitutes transmission of a thought from himself to his horse. This is why, as much as possible, horses should not change riders. The fusion that occurs between horse and rider is a strange phenomenon; in a sense the Greeks were being accurate when they created the centaur as a hybrid of man and horse.

Without doubt, the saddest thing to have befallen the equine world is the riding school or the dude ranch. This is the clearest sign of the horse's degradation. There he is trained in a manner exactly opposite from what his nature demands. There he must not be sensitive to touch: The knee of the "cowboy" who rents him must not be allowed to communicate anything to him, or a catastrophe might result. Desensitized, walking in a circle, mechanically obeying the commands of the manager, or going over the same trail to the end of his days, he has become a "carousel horse"—a discredit to man whom he has served so long and so well.

Unlike many animals, the horse has no genetic memory for training. He is born wild; succeeding generations of colts have to be taught everything from scratch.

Catching him, making him obey, and having him accept first a blanket, then a saddle, and finally a man on his back constitute the beginnings of his instruction. This is followed by the training itself, which varies according to whether one wants to make him a draft horse or a riding horse.

The young horse is gentle. He asks only to learn. This is not the case with the zebra, his first cousin! The zebra is actually untrainable, except when it comes to begging sugar from car windows in preserves where this dummy thinks he is living in the wild.

The horse cannot stand solitude; it is as though he retains the genetic memory of the herd animal. He becomes restless and develops neuroses, aerophagia (like society women), manias, tics, etc. To set him at ease he must be given a companion but definitely not a conspecific, with whom he might fight. In olden times a pinscher—a miniature Doberman—was assigned the task of ridding the stables of mice, and he willingly fraternized with the horse. But today he has become an indoor dog; he no longer wants to visit stables.

The sheep and goat have replaced this dog. Strange friendships have therefore developed. These relationships are sometimes very important to the horse: A triple crown winner may refuse to run if he does not have his friend near him up to the last minute . . . to cheer him on?

SEVENTEEN
Childhood

Slowly, gently, and without effort, the huge Indian elephant took the end of an enormous piece of wood in his trunk. At the other end his partner was doing the same thing. The undertaking seemed to be balanced, orchestrated like a ballet.

A few hundred yards farther on, the two elephants placed the gigantic log next to other logs which they had just transported. Then they lifted their trunks in a sign of joy: Their work was finished for the day. It was time for rest.

But while the male went to rejoin the other pachyderms and the mahouts, the female trumpeted briefly and headed at a leisurely pace in the opposite direction—toward the jungle.

None of the Indians present made any movement. All that happened was that her mahout smiled, tenderly and with complicity: the female was pubescent. She was leaving for the jungle in order to meet the wild elephant who would be her male.

A trained female elephant never allows herself to be inseminated by her work companion, nor will she return to stay in the jungle. It is as though a special relationship—a little like that which joins twins mysteriously—unites her with her mahout. She always returns to the man.

Three weeks later, at the hour when work was about to

begin in the lumber camp, the elephant returned and quietly took her place again.

A little later, the mahout had a baby, a son. The elephant herself was pregnant. During the course of the following 21 months, she saw her work time reduced progressively, until she was totaly exempted for the last six months.

But she stayed with the men and accompanied them on their work. In turn, they showed their affection by giving her tidbits of foods of which she was especially fond. As she had trouble rolling on the ground after a bath, they rubbed and curried her—and it was obvious that she was grateful to them for this.

Early one evening the elephant left again for the jungle. The men were therefore sure that she was about to give birth—it would happen in two days at the latest. This time she was not alone: Another female accompanied her to serve as midwife!

A week later the two pachyderms returned. Between these heavy masses trotted a tiny (just about 300 pounds) elephant calf.

The mahout's child was almost two years old. Under the benevolent eye of their parents—human and elephant—the two babies were introduced to each other.

During the 21 months that nursing lasted, the mahouts took care of the elephant, thereby respecting socialist principles even though ignorant of them! The mother hardly worked at all and could devote herself to the initial education of her nursling.

When the calf was three months old, the mahout's child was placed on his back; hands and trunks kept the two young creatures in balance. From that moment on, the boy was placed on his friend's back every day, first for one hour and then longer and longer. This was great fun for them. It lasted until the calf had reached the age of three years.

The two youngsters were hardly ever apart. They were so

used to each other that the calf trumpeted in despair when the boy left him for the night. But he trumpeted with joy the next day when the boy returned.

The Indian youngster was five years old. At this point "school" began in earnest for him and the animal: seven hours of lessons each day. Always together, they learned their common trade. By the time the boy was 14, each was as well brought up as the other, and they entered their "apprenticeship," rendering themselves useful by performing minor tasks.

Five years later both of them were earning their living normally. They spent five days a week at work, and two at rest. And they had three months of vacation.

As a factory worker forgets his work at vacation time and seeks out the tranquility of the country, a trained elephant leaves for his native jungle. Then, just like the worker, he returns to work when his vacation is over! It is not a matter of conditioning: The work he does he agrees to freely. It is an act of devotion. The man to whom he returns is his brother.

This is the nicest love story there is to tell about people and animals.

My son Tristan is a little blond boy of five years—perfectly normal, cheerful, and happy. Even though he is an only child, an entire family lives and plays with him: two cats, a dachshund, and an Afghan, Ramses, that is known to his intimates as "Big Yellow."

Recently Tristan saw a bitch get inseminated. It was very interesting but as normal for him as seeing his dog lift a leg. There was no problem: no nervous laughter, no embarrassment.

He has also watched a cat give birth. Very excited, he ran over to his mother and exclaimed, "It was so pretty when the baby came out of his mama's tummy!" He did not even ask her if he too had been born this way; this was beyond his imagination.

Thirty, fifty, or one hundred years ago, when adults would have blushed with shame at the mere mention of the word "sex" in front of a child, rural children knew as much about it as this little twentieth-century boy, and were no more shocked than he. This nature lesson is much more valuable than all those discussed by chairbound pedagogues.

In spite of this, there are still parents who do not want to have an "animal" because they "have a child." This is a terrific excuse—a dog will bite him, a cat will scratch him or, even worse, smother him, and so forth. Stories of babies found dead in the cradle with a jealous cat lying on top of them are as truthful as stories of vampires.

Is any of this true? None of it! Or, rather, one thing: a parental psychosis exists which definitely traumatizes the child about this subject.

"Stay away from the dog—he'll bite you." Or, "The cat will scratch you."

And if the child gets himself bitten or scratched one day, "I told you so!" exclaim the parents, who will themselves have inculcated in the child the fear that was the precipitating factor. Rendered timorous, nervous, and worried, emitting an odor that makes the animal aggressive, the child will never approach him except to touch his back furtively and then run away, or to throw stones at him.

But the animal will be held responsible, of course. Recently we have seen a typical case:

Monique was a little girl of seven years. Her parents were well-off artisans who lived in a tiny village. These people's hatred for animals bordered on obsession.

Across the way lived some farmers who raised their children and animals together. One morning their canary—the youngest child's pet—escaped from his cage and flew toward the yard of Monique's parents. The little boy knocked politely at their gate.

"Is my canary here?"

"Yes," said the father. "He is hiding in a corner of the yard. Come on in; you'll be able to catch him."

But just as the child bent down to pick up the bird, the man's big boot came down, crushing the canary. Monique, who witnessed this scene, snickered and went skipping off: "Hooray! Hooray!"

This village is perhaps the only one in France where one does not see any dogs, or cats, or goats, or geese in the street.

Monique's parents went to find the mayor: They felt justified in reminding him of the law concerning stray animals! The mayor shrugged his shoulders, but what could he do? He was obligated to enforce the law. But when Monique sat on a wall and threw stones at some dogs who, thanks to her, were penned in the yard, the law did not apply. Nor did it prevent her father from pouring cement in the nests of swallows!

One day a drama was acted out. A peaceable, self-respecting mutt, about a foot high, was the cause of it. Smelling a bitch in heat, the dog squirmed through the fence around the yard where he lived and suddenly found himself in this street where no animal ever came. As luck would have it, Monique came along at that moment and saw him (minding his own business, he did not even notice her).

Screaming in terror before this monster of the apocalypse, Monique threw herself to the ground, victim of a nervous crisis. The father came out, gun in hand, and fired at the amorous little dog. "He bit my daughter!" And the ancient cry echoed from the distant past: "He has rabies!"

He repeated this before the tribunal which considered the matter. In vain the judge and the veterinarian, who had autopsied the animal, tried to explain to him that the dog had been perfectly healthy. He brandished a medical document to the effect that Monique had been so "frightened" that she had had to remain in bed for six days!

But he took care not to repeat what the doctor had told him: that his daughter was not normal and ought to be seen by a psychiatrist. For if the animal is not at fault, then who is? Certainly not he himself—he is perfectly normal, is he not?

In spite of his unfair condemnation—two weeks' suspended sentence—he certainly did not regret having killed that dog! Not at all! And if the same situation arose again. . . .

A veterinarian, Dr. Condoret, has studied the interactions of children and animals for several years. He has observed the relations of 900 boys and girls from six to 13 years with "their" pets. According to his conclusions, the animal is the child's best friend, an excellent source of stability.

This is also the opinion of pediatricians. In most cases the presence of an animal reflects a basic physiological and psychological need in children. It enhances their comprehension of life and, above all, renders its mysteries "normal"—sexual behavior being the prime example. It also helps children become sociable, to become aware of bilateral exchanges, and not to cut themselves off with the blind egoism of "nidicolous" children, who tend to regard others solely in relation to themselves.

Nevertheless, it would be contrary to all proper education to give the child a pet as you would a toy, an object with which he can do as he pleases and which he can treat merely as a source of amusement. The act of "taking charge of" an animal, of experiencing a sense of responsibility, will teach him much more than any paternal sermon. For this reason a dog or cat is preferable to a fish or hamster. He will know that he has before him a living being, capable of suffering, affection, and also aggression, and thus having faults and good points like himself.

All pediatricians can cite cases of children who have stopped wetting their beds once they have house trained their dog. This is of practical value to the parent but is of even more importance for a child. For an "only" child, an animal can be the companion for whom he has an essential need.

If one were to tell Monique's father that in psychiatry the

animal can serve as a link to the world for seriously disturbed children who refuse all contact with the outside, he would not believe it. Nevertheless, in a case like that of his little girl, it is obvious that to avoid aggravating her very real—not just incipient—neurosis, she would have to be separated from her parents. Their own "brutish" cruelty was responsible for her problem. And to lead her back to a normal life, one would have to resort to the help of an animal!

EIGHTEEN
The Pet and "His" Child

"Big Yellow" was three months old when he entered Tristan's life; Tristan was then six months. Immediately they both fell victim to "puppy love." More than once were they discovered sleeping side by side, hands and paws intertwined.

As an adult, the Afghan tolerated all the ill treatment that the baby inflicted on him: fingers in his ears and eyes, little hands grabbing onto his long, silky hair, and the use of his tail as a rattle. But it was obvious that the dog considered himself, now a year old, to be dominant over the little "four-footed" youngster. For this reason the dog pushed the boy away gently when he had had enough of these games. Ignoring the baby's prattle (which was an attempt at calling the dog), Big Yellow very clearly watched over the baby, taking the latter under his protection, as is normal. Thus, when Tristan was in his playpen, the dog stood guard and forced him to stay there, knowing that the baby was not supposed to go out.

The pack, then, was arranged as follows: The father was the leader, or dominant male; the dog was the subordinate male; and the child was last.

Now, at about two years old, the child began walking. Standing upright on his hind legs, he assumed a dominant posture, which was inconsistent with his hierarchical position. This disturbed the social order and threw everything into question.

Baffled, the Afghan believed that his little brother had suddenly become dominant over him. Unable to understand this, the dog retreated to a corner and sulked. He had a good deal of trouble accepting this transformation from a four-footed pack member to a two-legged one. For a long time the Afghan rejected the little boy, fleeing even when the latter wanted to play with him. Two or three times the dog even growled in exasperation. This was normal; to stand upright is a dominant position. If a chimpanzee wishes to become the leader, he stands up on his hind legs and charges, waving his arms; normally he walks on his hands and feet.

Now that Tristan is five years old, Big Yellow has resigned himself to being dominated. The society is therefore hierarchically organized as follows: father, child, dog. But the dog, being motivated by parental impulses, is both the child's subordinate and his protector.

In almost all cases, when a child begins to walk, his pet faces a psychological crisis. The animal believes that his "offspring" is trying to dominate him and therefore he wants to make the youngster fall in order to reestablish the normal hierarchy. So it is best to be wary of the animal's behavior during this delicate period.

For this reason, too, it is best to buy a puppy when the child is already upright. One thus avoids having the animal traumatized by a perplexing change that seems unfair to him. The optimal time for adoption, therefore, depends on the child's age: between three and five years. During this period the upright position will be immediately accepted by the animal. Moreover, when the pet dies his master will be between 18 and 20 years old and will therefore sustain the loss better.

But if the animal—dog or cat—already is a member of the family when the baby is born, complications arise which resemble those that an older sibling experiences under the same circumstances. Either the animal will immediately be overjoyed by his "baby brother"—and will regress a bit, which is not very serious—or he will become jealous. The latter possibility is the more serious. The same precautions will have to be taken as with a first child: showing him plenty of affection—even more than before—and showing him the baby shortly after birth and patting both of them simultaneously. Problems will usually dissipate sooner or later.

A journal devoted to the study of social behavior in dogs, cats, and other domestic animals, has described two case histories that illustrate these points well.

The first is the story of Bouba, an 11-month-old Siamese cat, and Antoine, a 6-year-old child—the equivalent age. The parents introduced them to each other. Bouba, "very dignified and even a bit supercilious, pretended not to be interested in the matter...." As for Antoine,

> Upon seeing the cat he remained transfixed, stupefied. Then, stepping very quietly, he approached the newcomer and, without paying any attention to us any longer, he began to talk to him tenderly.... His fluted little voice became sweet and his usual brusque and clumsy movements were attenuated and full of gentleness. While still talking, Antoine extended his hand and began to pet the cat who, even though we had thought him rather wild, did not recoil at all. Lightly at first, and then more and more boldly and precisely, he scratched him under the chin, having always known that this was a cat's weak point. Just then Bouba, who had remained passive to that point, gave to Antoine a look of total beatitude. That was the beginning of a long love story. Their devotion to one another is still, after six years, just as intense and deep as it was then.

The other story, related by the psychologist Genevieve Jurgensten, concerns a one-sided love between her daughter and Triton the dog:

> When my daughter was born, Triton was already with us. The dog reacted, like a favorite child, with

jealousy toward the newborn who was coming to take his place in his parents' affections. From this moment on, he felt himself less important, rejected. He began to look depressed and defeated. Nothing interested him any longer, especially my daughter. On the other hand, the child immediately adored the animal. Each morning she first looked for Triton. She was rather peevish when she woke up, but as soon as she saw him, she began to smile and babble. He was her first source of interest; perhaps she identified with him. The dog's name was the first word that she pronounced: she had stammered "Trito" even before she said "Mama" or "Papa." Even now she looks for her companion constantly and tries to play with him. But all her advances are greeted with indifference by the dog. He is not cruel or even actually aggressive (one time he began to growl at the cradle, after which we scolded him so severely that he never did it again); he just did not like her. Nevertheless she did not hold it against him. I think she is genuinely fascinated by him and has a sincere affection for him. Maybe hers is not an ideal love capable of self-denial and great sacrifices, but she is heartbroken if the dog is sad, for example. Once I almost forgot about him when returning from the country and it was she who reminded me: She began to cry and shouted "Triton!"

It would be a serious mistake, especially in a case like this, to pay no more attention to the animal once the baby had arrived, or to relegate the animal to the kitchen, or to shout if he approached the cradle.

We have known one situation like this. A couple who had been married 12 years and were childless had adopted a dog early in their marriage. Then, when no one expected it, a baby came into the world. She was the joy of their lives. And of course the dog, 11 years old, was ignored completely. Having been a mere substitute for the baby whom they now had at last, he no longer interested them. He was in the situation of a doll that a little girl throws in a cupboard when she grows up.

But a dog is not a toy. This one stopped eating. His owners took him to the veterinarian (perhaps they felt guilty). The examination showed uremia, but so mild that there was no doubt that the dog would recover. Nevertheless the dog allowed himself—in the literal sense—to die.

And people maintain that animals do not commit suicide! In similar cases, the animal may go away, apparently preferring to wander about rather than to be neglected at home. This is why, if a couple feels that they will detach themselves totally from their pet as soon as their baby arrives, they should give the animal away before the birth. This will be much better for all concerned.

An animal cannot bear being replaced, from one day to the next, in his "parents'" affections. Nor can he understand why he no longer deserves a kind word or a caress—when everything is being lavished on a strange little larva! If the animal becomes irritable, or snaps at, growls at, or even bites the baby, whose fault is it?

On the other hand, if his masters continue to show him plenty of affection, he will have no reason to be jealous. He will embrace the child and, if he is old, will regain his vigor: He will have a "baby"!

"Dogs are not afraid of children; on the contrary, they show great confidence in them," Konrad Lorenz has said. Moreover, there are almost always heartwarming examples of extraordinary affection shown by the animal.

This astonishing devotion is explained in part by the parental behavior of canines. This is what makes a dog take to a child—especially a young child—and treat him as his own offspring. A recent hypothesis attributes the benevolence of the wolf and dog toward the young to a biochemical factor: It has been found that young mammals and babies have the same abundance of potassium and magnesium.

However, this chemical similarity would not appear sufficient to explain the dog or cat's enthusiasm for the child. Other affinities are present, the principal and best known of these being play and language.

Play is just as basic to animals as to children. It is certainly basic to the powerful link between them. An adult will play for 15 minutes, 30 minutes, or an hour, but he always stops: A child will play as long as the animal does. Often the child and his pet will fall asleep against each other—which is proof that both of them seek out the other's odor and contact.

But there is something else operating, and here we enter the still poorly understood world of language. It seems as though children and animals share a common language—half oral, half gestural—which they use as a code, it being undecipherable by adults.

It is known that animals understand each other very well and can exchange concrete information. Babies, too (from 1½ to 4 years old), employ postures and gestures that may qualify as displays. Thus, they express appeasement, offering, refusal, threat, and aggression—exactly as animals do.

The work of Professor R. Montagné suggests that these two "languages" might be readily comprehensible to animals and children, since certain of these gestures are originally common to both. Adults, on the other hand, succeed in understanding some of these gestures only by observation and deduction.

Two babies find themselves together in a playpen. They approach each other, but openly, with their palms turned downward in a sign of nonaggressiveness. Next a dog is introduced into the group. He understands the message very easily and responds to it by wagging his tail!

More astonishing still, it has been determined that the cries of hunger or pain of the young child, the puppy, and the kitten are emitted in the same very narrow register. This is a perfect example of sharing a code.

The apartments where Tiberius, a four-year-old German shepherd, and Marc, 18 months old, lived were on the same floor. Tiberius's owners and Marc's parents did everything they could to prevent the two of them from meeting: such a huge dog, and such a tiny baby!

But one day the two apartment doors were left open accidentally, and Tiberius was found in the other apartment, wagging his tail, with the baby nestled between his paws. Both of them were delighted. Marc cried when he was separated from his new friend. Sulking, Tiberius was dragged back home craning his neck to keep his eyes on the child. From then on the two of them had an obsession: to rejoin each other.

Why had Tiberius, a tough, brutal fighter with other dogs, felt this spontaneous love for a frail little being? He had a comparable affection—but clearly less intense—for a cat and a five-month-old dachshund. He played with the dachshund for hours at a time, without hurting him, just as he did the baby. One is forced to conclude that smallness and vulnerability appealed to him. So too did play, where he showed a gentleness that he never exhibited with an adult human. All you had to do to be convinced of this was see the care he took not to jostle the baby and to avoid making sudden movements which might scare or hurt him.

Can this characteristic parental behavior be said to occur every time a dog comes in contact with a child? Well, no. The same Tiberius detested the baby's older sister (four years old). Whenever she approached he growled, got out of her way, and fled with his tail between his legs. Doubtless he was jealous.

Thus it is quite difficult to explain by science alone the affinity that an animal may have for a child. Despite all that we have said, this remains the least explicable kind of love, a "something" that irresistibly attracts the animal toward someone. But this mysterious attraction must exist in order for close relationships to arise between animals and children.

This feeling of attachment can lead to the most unexpected behaviors.

There was, for example, the cocker cited by Dr. Condoret. This dog was aggressive with all the members of the family but maintained a privileged relationship, compounded of patience, gentleness, and calmness, with a single one of the children. And this child was a mongoloid.

There there was the little fox terrier who pulled the little deaf-mute girl by the skirt to warn her whenever she was alone in the house and the doorbell rang. If any other member of the family was home, the animal showed no reaction to the bell.

An animal's love for a child can go so far as to involve risking its life for him. Saada, a tame fox, loved a little ten-year-old girl. (As we know, the red fox is the best father of all the animals.) He played endless games with her in which he would run and jump over her outstretched arm. Starting ten yards away, this red arrow would fly over the little girl and land way behind her without even grazing her.

What happened that day? The fox and the little girl were having fun as usual in the immense room which served as their playroom. Did the child make a false movement? Did Saada misjudge his jump? Whatever happened his leap propelled him straight toward his friend's face.

He was only about two yards from her now, a distance over which his momentum would carry him in a fraction of a second. He let out a yelp, leaped to the side with all his strength to avoid her, and slammed against a wall. He had not touched the child.... Stretched out on the floor stunned, he gazed at her and wagged his tail.

NINETEEN
The Child and the Animal

Both of them had dark brown hair. They had the same lively eyes, changing from teasing to sad from moment to moment. And they were together constantly, a six-year-old girl and her 18-month-old puppy, Finette.

When she had been offered the puppy when it was four months old, she had taken it in her arms, had held it tightly to her breast, and, without saying thank you or even smiling, had retreated to a corner. She had returned ten minutes later to announce, "Finette is hungry, she wants some milk."

"How do you know?" her mother had asked.

The child knitted her eyebrows and looked at her mother. "She told me!"

Her mother had smiled in amusement. She would have been surprised indeed to learn that Finette may actually have "told" the girl that she was hungry.

The girl was an only child, spoiled, but withdrawn and very unhappy at the misunderstanding between her parents, whose quarrels she dreaded. But she recovered her equilibrium thanks to her dog. At last she had someone to talk to, to confide in, and to confess: "Last night Mama and Papa were fighting again, and I got up to listen. You know, Finette, I get so frightened when Papa yells . . . so scared. . . ."

"What are you muttering?" asked Mama.

"Oh, nothing. I'm just telling stories to Finette."

Yes, she was telling "someone" what weighed so heavily on her mind and what she could discuss with no playmate. She would have been "ashamed" to reveal these arguments: Parents are supposed to be perfect. Still less could she talk to them about it, since it concerned them and since she knew instinctively that she should not get involved in it. So she said, "Finette is sad," in order to express—through her pet's becoming herself—her unmentionable sadness.

To be sure, a pet could not supply the calm, warm parental atmosphere which she, like all children, desired. But at least it allowed her not to suppress these taboo topics, not to shut them up within her. In externalizing her chagrin, she rid herself of it in part. It may seem minor, to console oneself by fondling a dog, but at this age it is enormous.

Why do children get so attached to their stuffed dog, or Teddy bear, or their bunny rabbit with the big ears? Why do they insist on sleeping with them? Obviously an infant feels pleasure from having something soft to rub against, a security object to protect him from the terrors of the night. But at around two or three years his interest apparently shifts. At this stage he perceives his toy to be a real animal: What he likes about his dachshund doll is its resemblance to the real dog that he instinctively would like to have. It is as though prehistoric times, when the dog was man's indispensable partner, have left their imprint on the child.

It is at this stage, when animals begin to interest and intrigue the child, that family sentiment "pro" or "con" animals is set in motion. This can range from indifference (but it is not unheard of for parents to later yield on the issue of hamsters) to hate (Monique's parents).

In the latter case the child's behavior will necessarily be affected: He will be afraid of being bitten, or of diseases which the animal may transmit. Uneasy and fearful, always ready to hide behind his parents, he can easily become hostile, not only toward animals but also toward other children. He runs the risk of becoming a withdrawn, self-centered child. In serious cases this can lead to neuroses (Monique) or phobias.

This type of child, moreover, will always provoke a bite or a scratch because of his fear; this, of course, reinforces the parents' attitude. Even if he pretends not to be afraid, he will "exude" fear and the animal, "sensing" this, will attack. A normal child who sees an animal—no matter how large—is not afraid.

If the parents talk about animals in affectionate terms— even if they don't have any pets themselves—and are not afraid of animals, they will not inculcate fear in their child. They will teach the child to like animals and, hopefully, will find it natural, even useful, to take in an animal to share its life.

Animals appear in stories, pictures, rhymes, songs—in all of children's folklore. Often, as in Walt Disney cartoons, they assume anthropomorphic form and speak ... but this never seems to surprise children! These cultural phenomena confirm the important role of animals in the world of children, and the totemic value that they carry there.

As soon as he becomes aware of the fact that he is confronting a live animal, the child proceeds from discovery to discovery. He learns about its responses, becomes interested in its life, understands it, and "listens to" it.

Although he appears to learn his native language readily, the child actually finds the phenomenon of language very difficult to master. He prefers to express himself by gestures or by onomatopoeia, i.e., grunts or inarticulate sounds, like an animal. Therefore he has little meaningful contact with his parents for the first few years.

For a time, then, the baby and the animal learn the language of the family together. During the same period the child, with his blank memory, also discovers animal language. This is why, with the embryonic spoken language at his disposal, he is in a position to mediate between his parents and the animal. This relationship will persist so long as his gesticulatory language has not faded completely.

In return, the child will ask the animal to be interested in him—and not just for his petting or games. He wants to find, in his four-legged pal, a storybook hero, since these tales attribute human qualities to animals. This is the beginning of anthropomorphism.

The child also wishes to project his own desires, needs, and unsatisfied sentiments on his pet, to make the animal carry the burden of everything that worries, troubles, or wounds him. Especially in the early years, the child will ask the animal to be his double. Moreover, he will make the animal his defender—especially if it is a large dog—or, contrariwise, will regard his pet as a weak thing calling for protection.

For all these reasons it is good to let the child choose his own pet himself: its breed, but also its size, sex, coat, etc. He should be left to make his own choice, just as he does at school in making friends with one but not another of his playmates.

Thus children and animals develop in very similar ways. In particular, the resemblance in behavior between a young chimpanzee and a child is often striking. For example, when a small child meets another youngster, he readily touches the other's chin, thus expressing a desire to be accepted. For the same reason primates make exactly the same gesture!

On the other hand, the baby has to learn to speak in order to communicate with his parents since they have forgotten the language of gestures. This is why, incidentally, a "communication barrier" may occur between him and adults for two or three years.

It has been observed that it is when the child begins school that he is most likely to ask for a pet. It is as though his new scholarly role, which requires him to repress gestural language in favor of speech, accentuates his need for an animal—especially a dog—with which he can maintain his old way of communicating. For, as professor of psychophysiology R. Montaigné has observed,

> The young child who has feelings that he cannot yet describe by words can transmit them to an animal by using his whole body.... The dog understands the child and the child understands the dog.

This is why, if a boy or girl insists on having an animal, one should not refuse if there is no major obstacle. Behind this desire others may lie hidden, namely to confide in someone and to gain affection which is lacking.

All types of experimental schools make use of animals. More and more, goldfish, canaries, and even tame rabbits are found in classrooms. Animals facilitate introduction to the elementary ideas of biology. But they also present another advantage: They help develop the pupil's sense of duty, of society. Finally, animals also remind him that, in the middle of our urban civilization, nature is still present.

Today the increasing number of household animals and hence of contacts between animals and children allow some startling behaviors to be discovered. In a center for psychotic children where animals have been introduced successfully, the little patients say (as they themselves explain):

"I wanted a dog because I was not loved."

"My dog is my brother."

This is why the mere presence of an animal can sometimes modify the behavior of a disturbed child.

The cat, suspiciously regarded by young parents, nevertheless is the animal most suitable for a baby. The infant sees only poorly but his tactile and olfactory senses are extremely well developed. What small child has not delightedly stuck his face in his mother's fur collar, or run his hands through it?

The soft, electric, odoriferous fur of the cat pleases the baby sensually. The cat's rubbing against him enchants him because it is a sign of well-being that he perceives as such.

And if there is some danger, it is not because, as is popularly believed, the feline may smother the child. Rather, the opposite may occur: A kitten's thoracic cage is very fragile, and when a little child presses against the kitten, he can easily smother it! So if you wish to raise a cat and a child together, it is best if the cat is already eight or ten months old—for its own sake.

A feline is as patient with a child as a dog is. We once knew a very independent cat who was quick to unsheath her claws. But the little girl did what she wished with her, putting doll's clothes on her, taking her to bed, taking her out in a stroller. The little animal always purred peacefully.

The cat does have one fault. He teaches children human feelings: love and jealousy. This feline, whose personality is so close to our own, often responds as we do: This cannot be much of a virtue!

In a Paris suburb, two boys—brothers ten and 12 years old—succeeded in taming a cat that, wandering about half-wild, had decided one fine day to live with them since she found their house, garden, and the boys themselves to her liking. Gumball—that is what they baptized her—adopted the family: an extraordinary sign of trust.

Some time later a second cat, several months old, appeared and was baptized Mischief because of her impulsive nature. She was to share in the affection that had been reserved to Gumball up to then.

For the first weeks these ladies hated each other, were jealous of each other, and "insulted" each other. But soon they were reconciled around a cradle: Gumball's first litter awakened young Mischief's maternal instinct. Mischief even helped with the kittens, putting them back in their basket whenever their mother was seized by a taste for the vagabond life and took off for a short while. Being an appreciative sort, the mother cat accepts at once anyone who shows this kind of care to her young. So the two felines managed to muddle through life together on fairly good terms.

A good year slipped by. After being burglarized, the family decided to get a dog. They got a cross between a boxer and a Belgian shepherd: a rowdy, four-month-old female named Linda, who took over the house. Behavior that she had tolerated toward another cat, Gumball no longer could abide. The children petted the upstart, played with her, and generally showed her the affection that Gumball had originally believed was reserved for her and that she had already had to share once before.

Gumball gave vent to her anger, but in vain because the dog remained the favorite. So she went away sadly. The family looked for her, called her, offered her milk, meat, chicken ... but nothing worked. Only the love that she still had for the children made her pass the threshold of what had been "her foyer" from time to time. She even stayed for dinner and slept over once, but she left when she saw that Linda the dog had taken her place and was firmly entrenched. How human felines are!

Three thousand years before Freud, Jung, the other psychoanalysts, the zoologists, and all other "ists," a poet sang of the love of the animal for its master. When Ulysses returned home, the only one who recognized him was his dog, with whom, as a child, he had gone to play on the mountain. And the only tear that the hero shed was not for his wife or his child or the state of his kingdom, but for his pet, the dying Argos.

TWENTY
Sick Animals

Today our knowledge, as limited as it is, allows us to draw up a balance sheet for the animal in relation to his "superior," man.

On the positive side, animals have a brain and an endocrine system. They think, they have intelligence (more or less), and they have a language. Moreover, a mysterious parapsychological gift has been bestowed upon them, just as it has on certain people.

On the negative side, animals are not creative, except in a rudimentary way in certain primates. And we do not believe that they have a sense of death.

Psychiatrists add that they cannot become insane. This would seem to constitute a good point, but zoologists regard it as a lack in relation to man, to whom this privilege is reserved.

The matter is not so simple, however. In a large volume *Animal Psychiatry* edited by A. Brion, a professor of veterinary medicine, and H. Ey, a psychiatrist, a number of prominent neurologists debate this question earnestly over the course of 600 pages. The upshot is that, even if we cannot say that an animal can be insane or a victim of nervous depression, at least he can show "abnormal behavior."

These, then, are the subtleties and ethical issues with which scientists struggle. The fact that one cannot state with certainty that an animal is insane depends principally on the elementary truth: He cannot describe his state by means of language. In

...e has begun to dislike my husband and is so aggressive ...him that a constant problem has resulted."

...e problem was obvious to the naked eye: It was revealed ...e man's bandaged hand! "That happened yesterday," he .... "I returned from my factory and, upon arriving, I ...d over my wife to kiss her...."

...seemed to him a clear-cut case of animal jealousy. But ...ouple was so visibly upset at having to part with the fox ...r that the veterinarian decided to try a drug that generally ...s quite well on difficult children and irritable old men.

...everal weeks passed without news of the couple or the ... The veterinarian met the friends who had sent these ...ts to him. Curious to know if the treatment had worked, ...sked them if they had any news.

"Oh, you have worked a true miracle," they replied. "The ... has become the gentlest sweetheart! He loves his master ...w. But Mrs. Z ... has just been jailed! Imagine this: During a ...l mental breakdown she tried to kill her husband! And it ...med such a peaceful household...."

Yes, it "seemed" so. Actually, the woman hated her husband. Her upbringing had taught her to be submissive, to ...ver allow her true feelings to rise to the surface. But her ...ent aggression was manifested by her dog, to such an extent ...at he acted as a sort of medium.

The woman "bit" the man through the mediation of the ...imal. But as soon as the fox terrier was treated and had ...turned to normal because of the medicine, he no longer ...rved as catalyst. The social barrier evaporated, and.... What ...rs. Z could not admit was that she had taken refuge in ...sanity!

This dog, rendered aggressive vis-à-vis his master solely by ...he "proper" behavior of his mistress, suggests that the problem was not exclusively neurological. It seems that his subconscious reacted like a television set receiving an image transmitted by a receiver, in this case the wife.

The case of the fox terrier was unusual. But there are

terms that are sometimes sibyllir
trists explain:

> Human psychiatry dates from tl
> were able to observe mental pat.
> specialized institutions. All our
> originated at least 150 years ag
> patients . . . were available for s
> erinary art has never had anytl
> Therefore animal psychiatry is o
> mera: Can anyone imagine anyt
> and superfluous than postulating
> plex" in a dog, "delusions of gr
> "nymphomania" in an elephant? T
> psychiatry depends on verbal
> which the physician becomes awar(
> is experiencing. . . .

Poor little elephant, she has no rig
way any human has. But anyway, how
clare a deaf-mute person insane? Thi:
point of view—one at least—which is
highly abstract arguments of his colle
recognize animal psychiatry by arguing t
be raving mad," Dr. Lanteri-Laura said, "
a very superficial notion of madness, as
reverse anthropomorphism. . . ." This psy(
"at one and the same time on the problen
animal insanity, and the basic relationshij
This is very profound because the one i:
cause of the other.

Recently a very distinguished middle-
refined, went to see a veterinarian with thei

"We are very unhappy," the lady expl.
cause we are going to be forced to part wit
love very much; he has partially taken the
that we have not been able to have. But f(

simpler, almost everyday versions of it, especially in "city dogs," who unfortunately are more prone to mental imbalance than are farm dogs.

The unnatural life imposed by big cities suits dogs no better than it does people. Like his mistress or his master, the dog can have psychosomatic ulcers, hysterical paralyses, pseudopregnancies, etc. And, like man, he is treated with tranquilizers, sleep cures, and so forth.

Animals can also be subject to infant traumas capable of disturbing them for their entire lives, if no one comes to "psychoanalyze" them, the psychoanalyst then being the intermediary between the dog and his master.

We knew a poodle—calm, gentle, and affectionate—who, as soon as someone petted his neck, began to growl. At the same time he turned his head to stare hatefully at his own penis. As a baby he had had eczema in that sensitive spot. Whenever his master treated him for it, the man scratched his neck to calm him down. A typical Pavlovian reflex had given rise to a neurosis. It was very difficult to make it disappear.

Fortunately for them, animals have no notion of death. Therefore they cannot commit suicide. But it must be acknowledged that a dog may let himself starve to death if he loses his master. Cats, too, even though they are regarded as great egotists, often meet the same end.

Nowadays the majority of psychiatrists recognize this suicidal behavior and regard it as a behavioral abnormality similar to whatever prompts a person to shorten his days. But although an animal may stop eating because of despair, and may not struggle against either his depair or his weakness from lack of nutrition, he does not wish to die. Scientists have arrived at a clever explanation of this "false" suicide. Professor G. Deshaies concludes:

> Obviously this is not suicide in the sense that the term is generally used with regard to man. The animal does not let himself die knowingly and willingly. But a reac-

tive depression can be sufficiently profound to cause the death of the animal, and therefore is capable of inhibiting the instinct of self-preservation ... this is a sort of pathological grief reaction on the canine level.

You do not have to believe this, of course. To accept it is to be anthropomorphic, but still. . . .

What caused Finette, a German basset, to do what she did? Having her siesta tranquilly in the warm Algerian sun, at 3 o'clock on the dot she stood up and "howled and howled."

At the same hour, in Douai, a little town in the North of France, Finette's aged master suddenly died of a stroke.

This is not a story that someone told us. We were witnesses to it. In another incident I witnessed, horses belonging to the King of Morocco were stabled at Rabat. Yet one evening they perceived an earthquake at Agadir 350 miles distant, and were driven crazy all night.

One can always talk of coincidences. Some people suggest parapsychological phenomena. The inventor of the cathode ray tube, Sir William Crookes, spent part of his life studying parapsychological phenomena. Studied especially in the United States and Russia, parapsychology includes all psychic phenomena that are not understood as yet.

All the enigmatic facts that we have just related doubtless have the same origin as the "visions" of clairvoyants and the "wands" of witches. These are supernatural phenomena and it is understandable why psychiatrists are particularly interested in these cases, which are not just fairly frequent in animals but also occur in people.

Dr. Alain Assailly, a neuropsychiatrist, has studied these phenomena carefully on the human level. Under the rubric of "Assailly's syndrome" he has described the signs which can reveal a predisposition to these states in an individual. His research has suggested that all of the endocrine glands and the parasympathetic division undergo a great increase in activity which can lead to these supersensitivity crises.

Cats and dogs that get lost and then are guided by some mysterious sense to their master or their home represent a phenomenon that is no less baffling.

Thus Moune, a little short-haired domestic cat on vacation with her masters, was seized by an urge for adventure one day and escaped from their car. That was July, 1974, in Tarn. After having looked for the cat everywhere in vain, her masters returned home, to Mont-de-Marsan.

One day in February a plaintive meowing at their door caused them to exclaim, "It's not possible!" But there was Moune: thin, exhausted, and weak. She had spent seven months traveling 240 miles, but she had succeeded in finding her home.

More than 500 such cases have been studied at Duke University in North Carolina. The most interesting are those of animals that have succeeded in tracking their master under conditions that would seem to be impossible. One of the most astonishing, and also one of the best documented, had as its hero a dog named Prince. In 1914 his master had left England to fight on the French front. He had left his dog in London. One day he saw the dog rush up to him. No one ever found out how Prince had crossed the English Channel. But this is no more interesting than the other mystery: how he had located his master's scent among millions of others.

It is obvious that a domestic animal cannot exhibit the behavior of a wild animal. As intelligent, as "humanized" as he is, this excess of civilization causes him to suffer from disorders of his instinctual behaviors—neuroses. And it is quite possible that the delicate nature of his psyche is related to his "sixth sense."

These disorders, or imbalances, moreoever, can transform his behavior completely and therefore force him to reach a new equilibrium. The frog who wanted to be as large as a bull was indeed a megalomaniac.

But this does not mean that a cow who thinks she is a bison and inconsiderately charges the veterinarian who has come to take care of her has gone crazy (this happened in olden times and still does occur occasionally). Indeed, a surprising thing has happened to the cow, which has been a domestic animal for centuries, living in amity with man, accustomed to being milked, curried, and put out to pasture by his hand: Modern life, in cutting her off from the human environment, has restored some of her prehistoric instincts!

Machines have now replaced oxen, so the latter no longer know the yoke. And their sisters' udders, being handled mechanically, have forgotten the deft fingers of the farmer's wife. Crowded into herds in vast pastures, isolated from people, cattle have totally changed their way of life. The reconverted bull and the recycled cow no longer know their ancient friend, and the veterinarian is forced to treat them with a tranquilizer gun, just like wild animals.

On the other hand, in faraway Africa, the deer and wildebeest graze tranquilly near cars in their preserves. They now know that during the winter, no matter how cold it is, man will benevolently provide for their needs.

And for the fatherly lion that, as a cub, learned to play among the wheels of cars, the day will soon come when, nonchalantly ignoring gazelles as being too tiring to hunt, he will come to beg his daily steak from the hand of a Kenyan!